T0258734

Diabetes
A Comprehensive Treatise for Patients and Care Givers

Shamim I. Ahmad, BSc, MSc, PhD
School of Science and Technology
Nottingham Trent University
Nottingham, UK

Khalid Imam, MBBS, FCPS
Liaquat National Hospital and Medical Collage
Karachi, Pakistan

LANDES
BIOSCIENCE
AUSTIN, TEXAS
USA

VADEMECUM
Diabetes: A Comprehensive Treatise for Patients and Care Givers
LANDES BIOSCIENCE
Austin, Texas, USA

Please address all inquiries to the Publisher:
Landes Bioscience, 1806 Rio Grande, Austin, Texas 78701, USA
Phone: 512/ 637 6050; FAX: 512/ 637 6079

ISBN: 978-1-57059-775-6

Library of Congress Cataloging-in-Publication Data

Ahmad, Shamim I., author.
 Diabetes : a comprehensive treatise for patients and care givers / Shamim I. Ahmad,
Khalid Imam.
   p. ; cm.
 Includes bibliographical references and index.
 ISBN 978-1-57059-775-6
 I. Imam, Khalid, 1972- author. II. Title.
 [DNLM: 1. Diabetes Mellitus. WK 810]
 RC660.4
 616.4'62--dc23
                            2013045909

While the authors, editors, sponsor and publisher believe that drug selection and dosage and
the specifications and usage of equipment and devices, as set forth in this book, are in accord
with current recommendations and practice at the time of publication, they make no warranty,
expressed or implied, with respect to material described in this book. In view of the ongoing
research, equipment development, changes in governmental regulations and the rapid accumulation of information relating to the biomedical sciences, the reader is urged to carefully review and
evaluate the information provided herein.

Dedication

We dedicate this book to our children, parents, teachers, and all patients suffering from diabetes and their care-givers.

About the Authors...

SHAMIM I. AHMAD, after obtaining his Master's degree in Botany from Patna University, Bihar, India and his PhD in Molecular Genetics from Leicester University, England, joined Nottingham Polytechnic as Grade 1 lecturer and was subsequently promoted to SL post. Ahmad served for about 37 years at Nottingham Trent University (formerly Nottingham Polytechnic) before taking an early retirement to spend his time writing books and conducting full-time research. For more than three decades he worked on different areas of biology including thymineless death in bacteria, genetic control of nucleotide catabolism, development of anti-AIDS drugs, control of microbial infection of burns, phages of thermophilic bacteria and microbial flora of Chernobyl after the nuclear accident. But his primary interest, which started 27 years ago, is DNA damage and repair, particularly near UV photolysis of biological compounds, production of reactive oxygen species and their implications on human health including skin cancer and xeroderma pigmentosum. He is also investigating photolysis of non-biological compounds such as 8-methoxypsoralen+UVA, mitomycin C, and nitrogen mustard and their importance in psoriasis treatment and in Fanconi anemia. In 2003 he received a prestigious "Asian Jewel Award" in Britain for "Excellence in Education". He is the Editor of *Molecular Mechanisms of Ataxia Telangiectasia* and *Molecular Mechanisms of Cockayne Syndrome*, published by Landes Bioscience. He also edited *Molecular Mechanisms of Fanconi Anemia, Molecular Mechanisms of Xeroderma Pigmentosum, Diseases of DNA Repair, Neurodegenerative Diseases* and *Diabetes: An Old Disease, a New Insight* published by Landes Bioscience and Springer Science+Business Media.

About the Authors...

DR. SYED KHALID IMAM is an Assistant Professor, Consultant Internist and Endocrinologist. He acquired Fellowship in Internal Medicine from College of Physicians and Surgeons Pakistan and trained as Clinical Fellow at Liquat National Hospital and Medical College, Karachi-Pakistan, one of the biggest tertiary care hospitals of the country running into a private sector.

He is affiliated with above mentioned institution since 1997 and got all his postgraduate training, professional growth and positions from this esteemed institution and also fulfilled responsibilities as Head of the Department of Diabetes and Endocrinology, Program Director of Internal Medicine Residency Training, Chairman and member of Ethical Committee in the same institution for several years.

He is a member of American College of Endocrinology, an executive member of Pakistan Endocrine Society (served the Society as General Secretary as well), and a member of an Executive Advisory Panel of International Foundation for Mother and Child Health (IFMCH).

He had several publications at national and international levels and participated in many conferences internationally as invited speakers. Obesity, Diabetes and Metabolic Syndrome are the areas of his special interest and research. His future publications as co-editor include diseases of thyroid and obesity. He is also an author of three chapters in the book *Diabetes: An Old Disease, a New Insight* published by Landes Bioscience and Springer Science+Business Media.

Contents

Preface

The 21st Century is witnessing a tremendous resurgence of interest in and understanding of pathophysiology and genetics of diabetes mellitus and obesity, truly a pandemic of this century and causing a significant degree of morbidity and mortality due to an enormous increase in the incidence of cardiovascular diseases.

Diabetes is a complex disease and is also one of the most common of all diseases. It is almost impossible to reach to an accurate global prevalence of this disease as the standards and methods of data collection vary widely in different parts of the world. Additionally in many third world countries the patients are treated at private clinics and they seldom keep a record of patients specially those suffering from chronic diseases such as diabetes. Moreover, it has been shown that up to 50% putative diabetics are not diagnosed for up to 10 years. So at any given time we only account 50% of the diabetics than really exist. Nevertheless, a global estimate suggests that over 285 million people (amounting to 6.4% of the adult population) are suffering from this disease. The proportion of diabetes sufferer in different countries is also highly variable which depends on a number of factors including the genetic make up, life style and the environment they are exposed to.

Globally, diabetes, especially Type 2, is increasing with an alarming rate which is causing great concern not only for the general population, but also for internists, diabetologists and health educationists, and is imparting a heavy burden on health services. Three main reasons for the increased incidence of diabetes are a general increase in life expectancy leading to an increase in the ageing population, life style and the global rise in obesity.

Diabetes: A Comprehensive Treatise for Patients and Care Givers is written with the intention to keep readers abreast of latest advancement, understanding, emerging trends and technology in the field of diabetes. Several tables help to facilitate understanding of the concepts presented. Gestational diabetes, a topic commonly ignored, is also discussed in detail as a separate chapter.

This book also reviews recent findings of most popular herbal medicines to treat diabetes through their relevant mechanism of actions. The book is unique in the sense that it is written for both care providers and patients. Chapter 18, *Food and Diabetes*, the Epilogue and Appendices 1-4 are the sections esspecially designed and written for patients and care-givers.

An overview of above subjects in the field of diabetes should provide a solid background, foundation and information from which to view the exciting future developments in this rapidly moving science.

Shamim I. Ahmad, BSc, MSc, PhD
Khalid Imam, MBBS, FCPS

Acknowledgment

The authors of this book are cordially thankful and acknowledge Cynthia Conomos and Celeste Carlton for their guidance, patience and hard work in the preparation of this book. Also our thanks and acknowledgement goes to Ron Landes for his encouragement and support to bring the idea to produce the book. Finally the author (SIA) acknowledges Nottingham Trent University, Nottingham, UK for their support in the preparation of this book and the author (SKI) acknowledges support provided by Al-Mouwasat Hospital, Jubail Industrial City, Saudi Arabia.

CHAPTER 1

Introduction

History of Diabetes

For two thousand-years diabetes has been recognized as a deadly disease. The first recorded documentation of diabetes was by the Egyptian physician Hesy-Ra in 1552 BC. He characterized diabetes causing weight loss and more frequent urination (Polyurea). In the first century AD, Greek physician, Aretaeus used the term diabetes mellitus in which diabetes means "to pass through" and mellitus, a Latin word for honey (referring to sweetness) noting that the urine would attract ants. Now this name diabetes mellitus or DM is medically adopted worldwide. The initially described cases are believed to be of Type 1 diabetes. In 400 or 500 AD the two Indian Physicians Sushruta and Charaka were the first to identify Type 1 and Type 2 diabetes as separate conditions with Type 1 linked to youth and Type 2 with overweight.

What Is Diabetes?

Diabetes is a disease of metabolic disorder of glucose homeostasis in which blood glucose levels become abnormally elevated and causes harmful effects to the organs. However, glucose only appears in urine when plasma glucose exceeds renal threshold for glucose absorption.

It is important to note the misconception that *diabetes is a result of consuming too much sugar*. It is not yet proven to be correct. However, eating too much sugar or sweets may cause obesity, an important risk factor for diabetes. There are two basic reasons of diabetes (i) Pancreatic Beta (β) cell dysfunction causing impairment in the production and/or function of insulin which plays key roles in controlling the level of glucose in blood and (ii) Insulin Resistance or development of resistance by the cells to take up insulin. Glucose plays a number of roles in our body and the most important role is to provide energy to our systems. Another factor contributing to diabetes development is the defect in incretins hormonal system. This hormonal system releases Glucagon-like peptides in response to oral glucose load and helps in regulating glucose levels in the blood.

Diabetes: A Comprehensive Treatise for Patients and Care Givers,
by Shamim I. Ahmad and Khalid Imam. ©2014 Landes Bioscience.

A persistently elevated level of blood glucose is associated with diabetes related complications and diabetic patients therefore are prone to a number of short and long term complications including kidney damage leading to kidney failure, eye problems which can lead to blindness, neuropathy leading to phsiological impairments such as loss of sensation etc. Macrovascular problems include ischemic heart disease leading to heart failure, cerebrovascular diseases or strokes and peripheral arterial diseases.

The high level of glucose in the blood stream leads to damage of the lining of the arteries and other blood vessels and enhances the process of atherosclerosis, a phenomenon leading to a compromised blood supply to the organs. If the coronary arteries are affected it can lead to heart attack. If the arteries to the brain are damaged it can lead to stroke and if the arteries supplying to kidneys are damaged this can lead to nephropathy or poor kidney functions. Small blood vessels are also damaged by diabetes and this can affect the eye sights or retinopathy. Diabetes can also lead to poor circulation of blood in feet and legs, and in extreme cases it can lead to gangrene and amputation. Diabetes can also result in various other ailments such as delayed wound healing, periodontal or teeth disease, erectile dysfunction, depression etc. Hence diabetes is an important cause of prolonged ill health and premature mortality with approximately 1 death every 10 seconds globally.

At a later stage in this book each diabetes-related complication will be described in full detail. The development of complications depends upon duration and control of diabetes. Better the glucose control slower is the development of diabetes complications. In fact the record shows that patients diagnosed to have Type 2 diabetes at their usual age can live over 25 years with this disease so long the diabetes is well controlled.

Diabetes: A Silent Killer

It is a fact that in many countries (for example in Asian Sub-continents), specially among people living in urban areas, *diabetes is not considered a disease*. This is because the disease does not show any apparent immediate effect on the overall physiology or immediate threat of death such as shown in heart failure, cancer and acute infections. Furthermore diabetes usually does not result in fever, headache, sickness, palpitations, infections etc. Therefore, diabetes is a silent killer through its complications and it causes about 5% of all deaths globally each year. This rate of mortality is rising and is expected to rise by more than 50% which means 10% of deaths globally within 10 years if no cure found or effective preventative measures imposed.

Diabetes Can Be a Blessing in Disguise

Diabetes can be "a blessing in disguise" because many non-diabetic persons do not take good care of their health in normal life especially as far as the eating habits and life styles are concerned. As a result the obesity and its related diseases especially diabetes is on increase. Once the diabetes is diagnosed, those patients (especially the health conscious) immediately change their life style and not only go for healthier diet but also prevent themselves from eating excessively. They also go on for exercise and, especially if obese, try to reduce weight which benefits them for other diseases as well. Also their doctors, as a precautionary measure, may prescribe them cholesterol lowering and other prophylactic drugs such as aspirin and keep a watch on diabetes related complications. Since coronary heart disease is one of the leading causes of deaths globally and diagnosis of diabetes leads patients, through doctors, to take more precautions against it—we call it *A Blessing in Disguise.*

CHAPTER 2

Epidemiology

Global Prevalence of Diabetes

Diabetes is one of the fastest growing non-communicable diseases throughout the world. It is very difficult to get hold of the accurate counts for its global prevalence as the methods of data collection is widely variable in different parts of the world. In addition many patients are not included in the count because according to an estimate around 50% of the cases remain undiagnosed for up to 10 years. However, according to an estimate for 2011, globally there were more than 380 million people (amounting to about 7% of the population) suffering from this disease. In Britain alone, the recent survey shows that, there are about 2.8 million people currently suffering from diabetes. It is also estimated that by about 2025 this population will reach to about 440 millions.

The frequency of diabetes varies in different part of the world and in different populations, for example the incidence in Asia (India, Pakistan and Bangladesh) is estimated at around 10-12% of the population, whereas in Europe it is about 4-5%. Genetic makeup and the era at which the key genetic mutations for diabetes occurred may be attributed to this variation. Even within the same country the variation has been noted; for example, the distribution of diabetes in different regions of the United Kingdom is different: England, 5.4%; Northern Ireland, 3.7%; Scotland, 4.1% and Wales 4.9%. One possible reason for the higher proportion of diabetics in England may be due to the higher proportions of migrants from countries with high prevalence of diabetes.

Percent data calculated from previous publication (Diabetes: an old disease, a new insight, Springer publication, 2012) show the highest prevalence of diabetes is in certain European countries (8.9%) followed by in the Middle East and North Africa (8.4%) followed by Central American countries (7.7%) followed by South East Asia and Western Pacific countries (6.4%) and lowest in African countries (3.2%). If we look for the possible reason(s) for this variation, it can be postulated that diet, activities (exercise etc.) and longevity in age may be the leading reasons for this variation.

Diabetes: A Comprehensive Treatise for Patients and Care Givers,
by Shamim I. Ahmad and Khalid Imam. ©2014 Landes Bioscience.

Out of several types, Type 2 diabetes (for detail see below) contributes to approximately 85-90% burden of all types. Increased life expectancy leading to increase in aging population and general rise in obesity are most likely two main reasons for the increase in Type 2 diabetes. Diabetes and obesity are considered to go hand in hand. Economic stability and technological progress add to the development of obesity in many parts of the world including Europe, USA and in oil producing countries, Kuwait, Saudi Arabia and United Arab Emirates. It is also recognized that above the age of 60 the diabetes is 10 times higher than the age between 20 and 39. Socio-economic progression tends population towards high energy intake and reduces energy expenditure which adds further load on the pathogenesis of diabetes.

Diabetes is the fifth leading cause of death in the UK and over 200,000 people suffer from diabetes-related complications every year. Approximately 80% of people with diabetes live in low and middle income countries.

Despite the fact that a huge amount of research is carried out in this field, we are still a long way away from understanding fully the mechanisms involved in diabetes (especially Type 2) development, and it still remains an incurable disease. The treatment mostly and basically given is to control blood glucose levels thus prolong the onset of diabetic complications as much as possible.

Diabetes Burdens on Healthcare

In many developed countries the healthcare expenditure on diabetes and diabetes related complications alone is now reaching up to 10% of total national healthcare budget. In middle and low income countries for example in Pakistan, direct cost of diabetes care is even much higher than the national health care budget.

The changes in the life pattern, long living expectancy and lack of improvement in health care are in part responsible for the astounding rise in the incidence of diabetes. Also diabetes at times is not considered a disease and hence can be taken reluctantly. If the trend will continue it is likely that countries across the globe will face an economical crisis in the burden of healthcare. Hence there are challenges in the successful treatment of diabetes because of personal and economic costs incurred in its therapy.

According to a report in BALANCE (Bimonthly magazine from DIABETES UK, July-August 2012) vast majority of people in England do not receive the regular checks and services recommended by National Institute of Health and Clinical Excellence (NICE). As a result an increase has occurred in diabetic related complications such as stroke, heart attack, and leg amputations. These complications accounts for 10% of the entire budget of National Health Service. The recommendation is that there is an urgent need for early intervention, better risk assessment, screening and, more importantly, educating people suffering from diabetes especially of Type 2.

In recent years increasing number of reports appearing that children under the age of 14 are also developing Type 2 diabetes amid the already increasing prevalence of Type 1 diabetes (see ref. 141). Thus there is a serious global problem linked to changes in social and cultural activities, increasing urbanization, dietary changes and reduced physical activities.

According to BALANCE diabetic patients in UK are hospitalized 1.5 to 3 times more often than people who do not have it. Incidence of non-traumatic amputations, blindness, and terminal stage renal disease occur in diabetic patients. To delay the progression of diabetic complications and improve the quality of life it is therefore important to enter in comprehensive diabetes care. Also the fact is that acquiring knowledge about diabetes is a highly important part of diabetes management, and even more important is to make the patients aware of this chronic disease. "*For a diabetic patient, knowledge and understanding are not a part of treatment - they are the treatment*".

CHAPTER 3

Risk Factors for Diabetes

Just like many other diseases it is almost impossible to predict who will suffer from diabetes and when. However, there are some factors that can increase the possibility of developing diabetes:

Obesity

Obesity, by World Health Organization before 1997, was accepted only as a health risk factor but since then it is recognized as a disease. This recognition is based on epidemiological data which show that there is a worldwide increase of diabetes due to obesity, the increase in health expenditure and progress in pathophysiological concepts.

Obesity and diabetes are now becoming a "global epidemic." In 2000 there were around 300 million obese adults worldwide and within 10 years this number gone to around 500 million and by 2015 is expected to reach 700 million.

Body Mass Index (BMI) and waist circumference are two variables that classify the obesity (see Tables 3.1A and 3.1B) is one criterion to determine obesity and if BMI is in excess of 30 kg/m^2 the person is considered to be obese. A BMI of >40 kg/m^2 is considered as extreme obesity. If the BMI is above 25 kg/m^2 the person may be carrying the risk developing diabetes. Other simple way to determine obesity is to measure the waist circumference.

Around 90% of Type 2 diabetic patients usually suffer from insulin resistance due to excessive storage of body fat especially in abdominal area. Insulin resistance has been known as the main reason caused by accumulation of fat cells interfering with the cellular receptor's activities in transporting the glucose into the cell. Precise mechanism by which obesity leads to insulin resistance is not fully understood but may be related to several biochemical factors, such as free fatty acids, adipokines, Leptin and other biomolecules (see Chapter 8). Free fatty acid, the physiologically important energy substrates, is the most important component involved in insulin resistance. The mechanism through which free fatty acid induce insulin resistance involves accumulation of triglycerides and activation of several serine/threonine kinase enzymes. The increased supply of free fatty

Diabetes: A Comprehensive Treatise for Patients and Care Givers,
by Shamim I. Ahmad and Khalid Imam. ©2014 Landes Bioscience.

Table 3.1A. Classification of obesity based on waist circumference

	Waist Circumference
Caucasians	Men ≥102 cm
	Women ≥88 cm
Asians	Men ≥90 cm
	Women ≥80 cm

3

Table 3.1B. Classification of obesity based on BMI

		Body Mass Index Kg/m² Weight in Kg/height in m²
Caucasians	Normal	18 to 24.9
	Overweight	25 to 29.9
	Grade 1 Obesity	30 to 34.9
	Grade 2 Obesity	35 to 39.9
	Morbid Obesity	>40
Asians	Normal	18 to 22.9
	Overweight	23 to 25.9
	Grade 1 Obesity	26 to 29.9
	Grade 2 Obesity	30 to 34.9
	Morbid Obesity	>35

acid causes insulin resistance in skeletal muscle and liver, which contributes to the development of diabetes.

Obesity and diabetes are associated with other constellation of syndrome which is known as the Metabolic Syndrome (see Table 3.2). Presence of the metabolic syndrome in diabetics significantly increases the morbidity and mortality risk.

Family History

If one or more persons in the family have suffered or is suffering from diabetes, especially of Type 2, it is possible that other family members may also develop it. However it is not necessary that if one person in the family has suffered diabetes then the entire family is running the risk of developing diabetes.

Table 3.2 Criteria of the Metabolic Syndrome

Adult Treatment Panel III (ATP III)
Waist circumference *M: >102 cm, **F: >88cm
BSF >100 mg/dL
B.p >130/85 mmHg
TG >150
HDL-C M: <40 mg/dL, F: <50 mg/dL
3 or more criteria required for the diagnosis of metabolic syndrome
* Asian cut off = 90 cm, ** Asian cut off = 80 cm.

According to a report in BALANCE (May-June 2012) although 85% of Type 1 diabetes occurs in people who do not have a close family member with the condition, the risk among close family is about 15 times higher than the general population. About Maturity Onset Diabetes in Young (MODY, a disease caused by gene mutation in any one of the 6 genes identified, see Chapter 4) it is said that there is 50% chance that the child will inherit the mutation and go on to develop MODY before he/she is 25 years of age whatever is their weight, life style and ethnic group.

It is estimated that there exists around 3000 genetic diseases and according to The Global Genes Project Estimate, there are some 350 million people worldwide currently affected with one or other kind of genetic diseases. These are those in which mutation in gene(s) has occurred, been identified and the possibility of its passing over to the off-springs exists—hence also called inheritable disease.

Basically there are two kinds of genetic diseases, (a) those which are carried from generation to generation and (b) mutation induced in fetus and can be inheritable or non-inheritable. It is more difficult to predict or calculate the inheritance pattern in the latter class of inheritable diseases. In order to understand diabetes "in the family" or an inheritable disease, a brief outline is given below on inheritable genetic diseases.

What Are Genetic Diseases?

Genetics and genetic diseases are vast and complicated subjects and it is almost impossible to cover them in its entirety in this book. So, for readers, we will try to present below a flavor of this subject. Those who wish to learn more about genes should consult the text books in genetics in this regard.

All our biochemical functions in life are run by very complex biochemical reactions and systems which are governed by around 30,000 functional genes in our chromosomes. To understand genetic diseases it is important first to understand what are the chromosomes and genes?

Chromosomes and Genes

In virtually every kind of our body cells and in each cell of each kind is found a specific type of pocket known as nucleus and in the nucleus are located chromosomes. Chromosomes are made up of four kinds of chemicals called deoxyribonucleotides, and hence there are four kinds of deoxyribonucleotides, present in the form of a string (known as DNA strand). Each chromosome has two DNA strands twisted around each other in the form of a helix and hence it is called double stranded DNA. These four kinds of deoxyribonucleotides are present in different sequences generating a large variety of DNA sequence.

Human cells contain 23 pairs of chromosomes (22 pairs of autosomes and one pair of sex chromosomes), giving a total of 46 per cell. Females have two copies of the X chromosome, while males have one X and one Y chromosome.

The male gamete, called the spermatozoan, is relatively motile and usually has a flagellum. The female gamete, called the ovum, is nonmotile and relatively large in comparison to the male gamete. When the haploid male and female gametes unite in a process called fertilization, they form what is called a zygote. The zygote is diploid, meaning that it contains two sets of chromosomes.

The male gametes or sperm cells in humans and other mammals are heterogametic and contain one of two types of sex chromosomes. They are either X or Y. The female gametes or eggs however, contain only the X sex chromosome and are homogametic. The sperm cell determines the sex of an individual in this case. If a sperm cell containing an X chromosome fertilizes an egg, the resulting zygote will be XX or female. If the sperm cell contains a Y chromosome, then the resulting zygote will be XY or male

The 22 autosomes and X and Y chromosomes in human cells have been studied and have been found to carry around 30,000 functional genes. A gene on the chromosome in fact is a section of double stranded DNA responsible for a specific function, which is to direct the synthesis of a protein/enzyme. At one end of the gene is usually located the sequence to specify the beginning of the gene and at the other end is the sequence to specify the termination of the gene or "terminus". To be noted that this is a very simplistic introduction of the gene and the chromosome and the readers are advised to read more from a standard book on Molecular Genetics.

Although the mechanism of gene activities is more complex, however, to make it easy to understand, it can be said that each gene is responsible to synthesize a specific enzyme/protein. The enzymes determines which biochemical pathway will it run and when and how. Most enzymes are responsible to catalyze/drive the biochemical reactions and each set of reaction results in the production of a biochemical product (the building and driving blocks of the life). A huge variety and number of reactions take place in our body are known as metabolic pathways—the driver of the life.

Occasionally one or more gene(s) may be mutated (or altered) and become semi-functional or non-functional—called mutated gene and this can be transferred to the children from the parents, alongside other properties. Person whose gene has been mutated may suffer from a specific disease for which mutation has occurred. But as said earlier gene and gene associated diseases are driven by very complex mechanisms and it is difficult to cover them all here.

Extensive studies have been carried out on genetics of diabetes and a number of mutated genes identified. These genes may have been inherited from parents to their off-springs and hence those children inherited to carry the defective gene(s) may suffer from diabetes and some may not, although may be carrying the mutant gene(s) known as carrier. The carriers of genetic mutation (silence gene) may, in future generation, be expressed leading to the disease induction.

According to human reproductive system the genes of both, male and female are involved to produce children and if only one partner is the carrier of mutated gene there is probability that some offspring (but not all) will suffer from diabetes depending if the mutated gene is dominant or recessive. Those who do not become diabetic still can carry the probability of transferring the disease to the second generation or even to the third generation. The other possibility is that a child may be born with diabetes but none of the two parents suffered diabetes, nor were they carriers of mutation. These are those children whose diabetes gene may have become mutated during their own development in embryo (although it is rare).

Thrifty Gene

A hypothesis was proposed by geneticist James V. Neel in 1962 in which it was said that thrifty genes enable individuals to efficiently collect and process food to deposit fat during periods of food abundance.

The evolutionary biologists agree with the hypothesis that the early human ancestors posessing thrifty gene(s) enabled them to store up fat during food shortage and those who possessed stored fat were most likely to survive famine in compare to those who did not. As the thrifty gene products enables individuals efficiently collect and process food to deposit fat during period of food abundance it would have been advantageous for hunters/gatherers population, specially child-bearing women, to allow them to fatten more quickly during the time of abundance.

In the modern developed world, however, we rarely face serious food shortage, which means the thrifty gene is no longer useful in such communities. A serious drawback of the retention of this thrifty gene is that the retainers face struggles with their weight gain. Even if they exercise and eat healthy diet, their bodies may store the fat as if in a famine. An important question arising is: then why we are still keeping the thrifty gene? The answer may be that the famine has not altogether been eradicated from the world and secondly that during the period of certain illness the body fat is required to compensate them if supply of nutrients from external source is not enough or metabolism is impaired.

Ethnicity

As mentioned in the prevalence section that not only there are significant variations between various regions of the world ranging from 3.2% in African to 8.9% in European countries but there are significant variations within the regions; for example in Mauritius it is showing 17% of diabetic population while in Mongolia it is only 1.3%; in Saudi Arabia showing 13.6% down to 2.5% in Yemen. No doubt, obesity can be an important factor for the country-wise variations, but another factor contributing to diabetes is the genetic make up in which genetic mutation in that group may have occurred much earlier in the history than other group, thus increasing the high number of diabetic patients in the population.

It would be interesting to collect the diabetes data amongst the ethnic population with high prevalence to diabetes migrated to the countries where the prevalence of diabetes amongst indigenous population is relatively low; such as people from African origin settled in US and South East Asians settled in UK.

In a report (BALANCE, January-February, 20102) it has been shown that the risk of Type 2 diabetes is higher among black and South Asian people, in that in the later group the risk is 4-6 times higher than the former group where it is 2-3 times higher. Furthermore, it was noted that differences in insulin resistance and blood glucose levels were clearly present in 9-10 years of age group and that childhood obesity, physical activity and social class were not the main cause of these differences. In another report it has been claimed that in Britain by age 80 twice as many British South

Asian Black Africans and Afro-Caribbeans develop diabetes compared with Europeans of the same age. Also it is estimated that in future about 50% of this population of age 80 will be suffering from diabetes. Higher body fat and increased resistance of insulin (and some yet unknown factors) may be the reason for this increase.

Gender

Studies have shown that men carry higher risk of developing diabetes than women. An important reason for this gender-based difference may be because men normally carry more fat, especially around their belly. It is interesting to note that there is misconception that bulge around belly is normal growing process amongst men—this idea must be suppressed to protect them from diabetic risk.

Another gender specific diabetes, known as gestational diabetes mellitus (GDM), specific to women (for detail see below), is caused due to glucose intolerance that is first detected during the pregnancy and is associated with a probable resolution after the end of the pregnancy. This type of diabetes is one of the commonest medical conditions affecting ~7% of all pregnancies. The prevalence may range from 1 to 14% of all pregnancies, depending on the population studied and the diagnostic tests employed.

Polycystic Ovary Syndrome

Polycystic ovary syndrome is a common problem (around 1 in 5) in women during their reproductive life. Among young women it is one of the commonest causes of infertility due to loss of ovulatory cycle.

Polycystic ovary syndrome increases the risk of diabetes especially of Type 2 and heart disease. Insulin resistance is the key pathophysiological mechanism linked to polycystic ovarian syndrome and diabetes. Therefore, insulin sensitizer such as metformin is recommended for its treatment.

Age

Age is an important risk factor for diabetes and people of 65 and above run relatively higher risk of developing Type 2 diabetes.

Prediabetes

Border line diabetes, also known as prediabetes, is the kind which falls between the non-diabetic and diabetic profiles; hence it is in the grey area of diabetes. Parameters for prediabetes are:

1. Fasting plasma glucose, 100-125 mg/dL (5.5-6.9 mmol/L)
2. Two hours postprandial plasma glucose, 140-199 mg/dL (7.7-11 mmol/L)
3. HbA1$_c$ 5.7 to 6.4%.

Ironically, people in this category are at the risk of developing Type 2 diabetes and also cardiovascular complications. Patients falling in this category are advised to change the life style which might reverse the prediabetes to non-diabetes category.

3

CHAPTER 4

Classification of Diabetes

A number of different types of diabetes have been identified but most commons are Type 1 and Type 2 and gestational diabetes. These are describes below.

Type 1 Diabetes

Type 1 diabetes, which earlier was known as insulin dependent diabetes, typically seen in younger age group but may also occur in elderly population. This definition of "insulin dependent" is no more valid as Type 2 diabetic patients at later stage of the disease may also become dependent on insulin.

Type 1 diabetes is an autoimmune disease which means that body's immune systems of diabetic patients misrecognize the β cells, in the islets of Langerhans, as foreign bodies and progressively destroy them with the help of CD4 and CD8 killer T cells and by macrophages infiltrating the pancreatic islet cells. The destruction of these insulin producing systems naturally significantly reduces or stops the insulin production and hence the Type 1 diabetic patients from very beginning become dependent on insulin to be administered in order to control diabetes.

The clinical manifestations of Type 1 diabetes have been well established, and insulin administration has long been the routine therapy. However, the precise immunologic, genetic, and physiologic events that control the disease initiation and its progression have yet to be fully elucidated.

Type 1 diabetes has been associated with many other autoimmune disorders such as celiac disease, autoimmune thyroid dysfunction, polyglandular autoimmune syndrome, autoimmune adrenal insufficiency, pernicious anemia etc.

Subtypes of Type 1 Diabetes

There are two subtypes of Type 1 diabetes (i) Autoimmune or Type 1A (90% of the cases) and (ii) idiopathic or Type 1B (10% of the cases). Recently a new subclass of Type 1 identified and termed Fulminant diabetes.

Diabetes: A Comprehensive Treatise for Patients and Care Givers,
by Shamim I. Ahmad and Khalid Imam. ©2014 Landes Bioscience.

Type 1A

Approximately one-third of the disease susceptibility is due to genetic reason and two-thirds to environmental factors. Full information on environmental factor responsible for the increased risk is not available. A number of different factors including infections with certain viruses such as rubella, Coxsackie B4, cow's milk and perinatal factors have been associated with this type of diabetes (see Chapter 9).

The speed with which the β cells are destroyed varies with different individuals and can be short lived such as in neonates or may be long lasting and can be confused with Type 2 diabetes, as seen in latent autoimmune diabetes in adults (LADA). The antibodies appear fairly early in the autoimmune process in the circulatory blood before significant β cells are destroyed and can be used as markers. Several antibodies identified but the best studied are antiglutamic acid decarboxylase (anti-GAD), IA-2, IA2-β and anti-insulin antibodies.

Type 1B

It is a subset of Type 1 diabetes which does not involve autoimmunity. It lacks HLA association and is prone to develop ketoacidosis. Hence they were called "ketoacidosis-prone diabetes. Exact mechanism for this type of diabetes is not fully worked out.

A minority (less than 10%) of diabetic patients show no evidence of pancreatic β cell autoimmunity to explain their insulinopenia and ketoacidosis. This subgroup has been classified as "idiopathic Type 1 diabetes mellitus" and designated as "Type 1B." A recent report shows that about 4% of the West Africans diabetic patients with ketoacidosis are homozygous for a mutation in *PAX-4 (Arg133Trp)*—a gene that is essential for the development of pancreatic islets.

Fulminant Type 1 Diabetes

Fulminant Type 1 diabetes is a subtype of Type 1B diabetes. A novel another subset of rare kind of Type 1 diabetes with a remarkably acute onset, characterized by its extremely rapid progressing and complete destruction of pancreatic β cells leading to almost no residual insulin secretion even just after the onset. This kind of diabetes is more common in East Asia, notably in Japan where it is speculated to be prevailing between 5,000–7,000 cases or approximately 20% of acute onset Type 1 diabetic patients or 0.61% of all types of diabetes receiving insulin therapies. The disease has also been reported from China, Korea and a few cases from European countries. In Korea the prevalence of Fulminant diabetes is 7.1% in newly diagnosed Type 1 diabetic patients. Male and females are equally susceptible to this disease. The mean age of onset is 35 years in females and 43 years in males.

A significant decrease of β cells, α cells and mononuclear cell infiltration in both endocrine and exocrine pancreas is the landmarks of this class of diabetes.

Laboratory tests have revealed that the patients at the onset of disease carry a near normal level of HbA1$_C$, ketoacidosis, elevation of serum pancreatic exocrine enzymes and absence of anti-islet auto antibodies such as antiglutamic acid decarboxylase (GAD) antibody or anti-insulin antibody.

Type 2 Diabetes

Type 2 diabetes is the most common of all types of diabetes (ranging from 80-85% of all kinds) and usually comes later in the life, commonly after the age of 40 but also seen sporadically in younger individuals. It also develops due to sedentary life style and consumption of high caloric food.

Type 2 diabetes is different from Type 1 in several respects (see Table 4.1).

Insulin resistance is the hall mark of Type 2 diabetes encountered in 90% of cases. β cell dysfunction, elevated glucagon (α cell dysfunction) and incretins also play roles in Type 2 diabetes. Many patients with Type 2 diabetes are obese and even those who are not very obese often have raised waist circumstances.

Table 4.1. Difference between Type 1 and Type 2 diabetes mellitus

Features	Type 1 DM	Type 2 DM
Age	Generally younger	Generally older
Physique	Thin	Obese or overweight
Acanthosis nigricans	Absent	May be present
Symptomatic	More often	Less often
Complications at diagnosis	Absent	Often present
Family History	Negative	Positive
Ketosis prone	More	Less
Autoimmune markers	Positive	Negative
HLA association	Positive	Negative
Treatment	Insulin	Oral agents with or without insulin

Progressive β cell failure is an important feature of the disease. As the impairment in insulin synthesis and its secretion is not immune mediated the disease usually does not reach to the point where the patient requires insulin for the control of diabetes.

Interestingly about 20% of the Type 2 patients are not obese. In order to answer the question a team from Medical Research Council of UK in recent year discovered a gene known as "Lean Gene" This gene is involved in inducing Type 2 diabetes and heart disease in lean persons. The gene *IRS1* is responsible to determine the body fat. It is also responsible to having unhealthy levels of cholesterol and glucose in the blood. The suggestion is that leanness is not an assured criterion of averting diabetes and heart diseases but being more active and eating healthy diet is most effective way to reduce the risks.

Other Specific Types

i. **Drug-induced diabetes.** Glucocorticoids, thyroid hormone, diazoxide, Beta-adrenergic agonists, thiazide diuretic, pentamidine etc can induce diabetes.

ii. **Diabetes due to mutation defects in β cell function.** In hepatocyte nuclear transcription factor (HNF) 4-a (MODY-1), glucokinase enzyme (MODY-2), hepatocyte nuclear transcription factor (HNF) 1-a (MODY-3), insulin promoter factor-1 (MODY-4), hepatocyte nuclear transcription factor (HNF) 1- (MODY-5), neuro D1 (MODY 6) and mitochondrial DNA.

iii. **Diabetes due to defects in insulin action.** Are Type-A insulin resistance, lipodystrophy syndrome, leprechaunism, Rabson-Mendenhall syndrome.

iv. **Diabetes due to endocrinopathies.** These include acromegaly, Cushing's syndrome, pheochromocytoma, glucagonoma and hyperthyroidism etc.

v. **Diabetes due to diseases of the exocrine pancreas.** These include pancreatitis, pancreatectomy, hemochromatosis, cystic fibrosis and neoplasm.

vi. **Genetic syndromes associated with diabetes.** A number of genetic syndromes have been associated with diabetes which includes Down's syndrome, Klinefelter's syndrome, Turner's syndrome, Wolfram's syndrome, Laurence-Moon-Biedl syndrome and Prader-Willi syndrome.

Gestational Diabetes

Gestational diabetes is different from other diabetes especially from Type 1 and Type 2 which are more common and long lasting. Gestational diabetes occurs during the pregnancy and resolves soon after delivery in majority of cases. This diabetes is one of the commonest medical conditions affecting as much as up to 70% of all pregnancies and this frequency varies depending on the population studied and the diagnostic tests employed. However, it is more common in certain ethnic group such as South Asians and Pima Indians.

Gestational diabetes can occur even though there is no prior history of diabetes in the family or history of abnormal glucose intolerance (prediabetes). However if there is history of diabetes in first degree relative, its probability is significantly increased. This diabetes can occur even in case of normal pregnancy weight and without any obstetrical outcome.

About 30% of gestational diabetic patients also run the risk of developing Type 2 diabetes at later stage in the life. For management of this kind of diabetes please see Chapter 16.

4

CHAPTER 5

Glucose Production and Its Control in Body

In order to understand the mechanism of diabetes induction it is important to understand first the various aspects of glucose production, the structure and function of pancreas, insulin synthesis and the regulation of glucose concentration in the human body specifically in the circulating blood.

Glucose is a chemical substance derived from carbohydrates. By definition a carbohydrate is a compound which contains carbon, hydrogen and oxygen. Different carbohydrates have different structure and different number of these three atoms. According to their structure carbohydrates have been classified into monosaccharide, disaccharide and polysaccharides. Monosaccharide, containing cyclically arranged 6 carbon, 12 hydrogen and 6 oxygen atoms ($C_6H_{12}O_6$) are also referred as sugars (important to note that all carbohydrates or sugar are not sweet). Besides glucose, galactose (found in milk) and fructose (found in fruits) are important monosaccharides. Sucrose and lactose are disaccharides and cellulose and starch belong to polysaccharide group of carbohydrates. Certain sugars such as sucrose, lactose, fructose etc, when taken as food or via fruits (containing these sugars), in the body various enzymes biochemically convert these sugars from one form to other.

Carbohydrates perform many important roles in our body. Certain polysaccharides such as starch (found in wheat and rice, barley, oat etc) and glycogen serve as the store house of energy and some others (cellulose and chitin) acts as structural components mostly found in plants and certain insects. Sucrose, the disaccharide are found in sugarcane and beet root. As said earlier that not all sugars are sweet, likewise not all sweet compounds are sugar; some sweeteners are chemically produced such as aspartame which is a conjugate of two amino acids, phenylalanine and aspartic acid.

Glucose is a vital constituent of our life as it is used to produce energy in the body; also brain exclusively needs continuous supply of glucose. Glucose mainly comes from carbohydrates we consume as foods (such as bread, potato, rice). Fruits and dairy products can also supply glucose

Diabetes: A Comprehensive Treatise for Patients and Care Givers,
by Shamim I. Ahmad and Khalid Imam. ©2014 Landes Bioscience.

and various kinds of sugars can be an important source of glucose. It is also produced by the liver. If we do not take sufficient glucose with our food our survival can be in jeopardy. Around 200 g of glucose is produced and utilized in a normal adult body each day. More than 90% of it comes from liver glycogen and hepatic gluconeogenesis and the rest from renal gluconeogenesis. The other 10% we consume from external sources. The brain utilizes most glucose. It requires about 1 mg/kg body weight per minute or roughly about 100 g/day for a body weight of 70 Kg. The glucose in brain is most essential and is insulin independent. Glucose used by brain is oxidized to carbon dioxide and water. Other body tissues such as muscle and fat are to some extent non-essential glucose consumers. Glucose that is taken up by the muscle cells is stored there as glycogen or at times broken down to lactate which later enters in blood circulation where it becomes a major source of glucose by gluconeogenesis. Fat tissues also utilize glucose as a source of energy and as a source of triglyceride synthesis. Triglyceride and glycerol are then released from the fatty acids by its lipolysis; another source of liver gluconeogenesis.

Polyol Pathway and Diabetes

Polyol pathway is involved in glucose breakdown but it does not work when blood glucose is normal (e.g., 5.5 mmol/L); this is because the affinity of glucose for the enzyme aldose reductase, driving the first step of the Polyol pathway reaction, is low. However, as soon as the level of glucose is up, this reversible reaction becomes operative and the first step of the reaction is the conversion of glucose to sorbitol by the enzyme aldose reductase. In this reaction glucose is converted to sorbitol and NADPH to $NADP^+$. In the next step sorbitol dehydrogenase oxidizes sorbitol to fructose, which produces NADH from NAD^+. Hexokinase then phosphorylates fructose to fructose-6-phosphate while returning to the glycolytic pathway.

The NADPH is an important biological component which acts to promote the synthesis of nitric oxide and glutathione. Hence its deficiency causes glutathione deficiency which is an important scavenger of reactive oxygen species (ROS). The increased level of ROS leads to oxidative stress which is a major contributor of damaging cellular systems including the most important component, the DNA.

In essence, in this pathway the enzyme, aldose reductase leads to formation of reduced glutathione and an increase in ROS. ROS are known to highly reactive elements and can cause multiple damages to various cellular components including nucleic acids (DNA and RNA), proteins and lipids.

In human body most cells are dependent on insulin for glucose uptake. Other cells such as retina, kidney and nervous tissues are insulin-independent, so glucose moves freely across their cell membranes. Thus in diabetic patients, the excessive induction of Polyol pathway increases intracellular and extra-cellular sorbitol concentrations, increased concentrations of ROS and decreased concentrations of nitric oxide and glutathione. ROS has been branded as a major source of diabetic complications specially retinopathy and neuropathy.

Advanced Glycation End Products (AGEs) and Diabetes

Advanced glycation end products (AGEs) are the result of long term accumulated glycated damage to various cellular molecules which have a low turnover and are not replaced regularly. This non-enzymatic process initially forms reversible early glycation products and later irreversible AGEs.

AGEs have been implicated in inducing ROS-induced oxidative damage to various cellular components including diabetic neuropathy and nephropathy. Both types of diabetes patients show a higher level of oxidative damage to their DNA and matrix proteins than normal healthy humans.

Vitamin E and vitamin C have shown some promises as the former vitamin has the ability to scavenge ROS in vitro and vitamin C may function as an aldose reductase inhibitor preventing the sugar being converted to their corresponding alcohols. However their importance in human health per se is still debatable.

One of the latest entries in the antioxidant arsenal for diabetics is α lipoic acid. It is a naturally occurring dithiol compound and an important cofactor for mitochondrial bioenergetic enzymes. Major benefit of supplementing this compound has been found in diabetic neuropathy.

CHAPTER 6

Pancreas, Its Structure and Function

Pancreas is a gland which plays several important roles in digestion and endocrine activities. As an endocrine gland it produces several important hormones including insulin, glucagon and somatostatin, and as a digestive gland it secretes pancreatic juice containing digestive enzymes. A number of enzymes synthesized by pancreas, in combination with other digestive enzymes assist the absorption of nutrients and the digestion of food especially in the small intestine.

Depending upon the function each pancreas has been divided in two sections: the endocrine and exocrine glands. The microscopic examination reveals two different kinds of gland-stained tissues: a light-stained cluster of cells (approximately one million in number) known as islets of Langerhans cells and other is dark-stained known as acinar cells. The former type of cells produces insulin, the important hormones regulating blood sugar. The pancreatic exocrine gland helps the digestive system. It secretes pancreatic based digestive enzymes which are passed on to the small intestine helping in to further break down the carbohydrates, proteins and lipids (fats) in the food pulp.

As said above the islet of Langerhans plays an important role in glucose metabolism and regulation of blood glucose concentration. In the islet there exist four types of cells which are difficult to be differentiated by stains, but can be identified by their ability to secrete various enzymes. The α cell secretes glucagon, β cell insulin, γ cell somatostatin (regulator of α and β cells is also involve in growth) and PP cells, pancreatic polypeptide. The biological role of PP cells is not yet clear.

Pancreatic β Cell and Its Function

Pancreatic β cells which make up 65-80% of the cells in the islets are highly specialized cells responsible for storing and releasing insulin. These cells play a major role in human health as any disturbance in its production or activities can lead to diabetes. At the time of release the stored insulin in β cells also keep on synthesizing insulin.

Diabetes: A Comprehensive Treatise for Patients and Care Givers,
by Shamim I. Ahmad and Khalid Imam. ©2014 Landes Bioscience.

Apart from insulin, β cells release C-peptide, a by- product of insulin synthesis, into the bloodstream. It is suggested that this peptide helps in prevention of neuropathy and other diabetes related vascular complications.

Islet amyloid polypeptide (amylin) is stored in insulin secretory granules in the pancreatic β cells. It is co-secreted with insulin and is present in increased amounts in the pancreas of many patients with Type 2 diabetes. In this type of diabetes this polypeptide aggregates to form amyloid fibrils that are toxic to β cells of pancreas.

United Kingdom Prospective Diabetes Study showed that at the time of diagnosis of Type 2 diabetes, usually there is already ~50% loss of β cell functional mass, which is followed by a further progressive decline over time. Accumulation of amylin derived amyloidosis may be the main reason of β cell death in Type 2 diabetes.

6 Insulin: Its Structure and Function

Insulin is one of the smaller proteins in human body. The gene for its synthesis is located at chromosome 11 and is composed of 51 amino acids. The 51 amino acids makes two polypeptides, the A chain of 21 amino acids and the B chain of 30 amino acids. The A and B chains are attached to each other by two disulfide bridges. Each disulfide bridge links cysteine to cysteine amino acid, located at 2 different places in the two chains. There is also a third disulfide bond that connects these same amino acids within Chain A. Insulin is initially produced as pre-pro-insulin, which is transformed into a pro-insulin molecule by the action of a proteolytic enzyme, and finally into the active polypeptide hormone, insulin.

After the insulin is produced in the pancreas it is bound to an extra molecule known as C-peptide. This molecule helps insulin to assemble itself in most correct form. This peptide is then released from the insulin molecule in the blood stream where it remains until reused.

Glucose plays an important role in our daily life and its imbalance induces a variety of ailments specially diabetes. When we consume carbohydrates (most staple food contains carbohydrates), various biochemical reactions in our body converts them into glucose and this is the main source of body energy. For the glucose to be utilized, before its conversion to energy, it first goes to the blood stream and then to various cells. Glucose to enter into the cells from the blood it gets help from insulin to do so.

As said above, the pancreas of diabetic patients either cannot produce insulin or produce in subnormal quantities. Insulin acts as chemical signaling element and its main role is to intricately control the amount of glucose in blood. Alternatively there may be enough insulin in the body but its entry into the cells is impaired and this is known as *insulin resistance.*

In either case levels of glucose especially in blood becomes abnormally elevated and causes diabetes.

Glucose-Induced Insulin Secretion

Insulin is secreted in response to glucose and occurs in two phases: in phase 1 insulin release begins within few minutes of stimulation, after which it declines. In phase 2 insulin secretion begins a few minutes after the first phase, and secretion gradually increases to a peak within 30-40 minutes. The overall rate of insulin secretion is controlled by the concentration of glucose to which the β cells are exposed, the interaction of glucose with glucokinase, and the physiological state of the β cells. Exact mechanism of insulin secretion from β cells, induced by glucose, is a complex process and is an important subject of recent research.

Once insulin is produced, triggered by the high levels of glucose, it plays three important roles, (i) helps to transport glucose from outside to inside of the cell, (ii) signals the liver to convert the excess glucose to glycogen for storage, and (iii) signals other body cells (adipose/skeletal muscle cells) to take up more glucose. The process is a complex mechanism and requires a whole lot of understanding of various biochemical processes.

Effect of Insulin on Glucose Uptake and Metabolism

At the first stage insulin binds to its receptor which leads to translocation of Glut-4 transporter to the plasma membrane and influx of glucose, glycogen synthesis, glycolysis and fatty acid synthesis leading to a cascade of protein activation.

The whole process results in bringing the blood glucose levels down to normal levels. If the concentration of circulating glucose level goes below the required levels the cell of pancreas are stimulated to release glucagon which signals liver to convert glycogen into glucose and be released into the blood. This is how the levels of blood glucose are maintained at the required levels and the system is known as glucose homeostasis.

CHAPTER 7

Pathophysiology of Diabetes

Insulin Resistance

Insulin resistance is the hall mark of Type 2 diabetes. As said earlier almost 90% of Type 2 diabetic patients suffer from insulin resistance and become hyperglycemic (increased glucose in blood), although there may be enough insulin in the plasma but cells fail to transport glucose from outside to inside of the cell. In non-diabetics especially the fat and muscle cells require insulin to absorb glucose. In diabetic patients insulin resistance in muscle and fat cells reduces glucose uptake. In liver cells insulin resistance reduces glycogen synthesis, its storage and a failure to suppress glucose production and release into the blood. Hence insulin resistance normally refers to reduced glucose-lowering effects of insulin. Exact mechanism for insulin resistance is not fully worked out as it is a highly complex process. Moreover, in the majority of cases with insulin resistance, obesity is almost invariably present and responsible for a rising trend of diabetes not only in adults but also in children.

Insulin resistance usually occurs due to impairment of insulin pathway at any steps. For example in human insulin resistance has been proposed to occur as a result of rare mutations of the IRS-1 protein, although most data on insulin resistant caused by IRS-1 protein have been obtained by animal experiments. Several other enzymes such as PI3-kinase, PKB/AKt2, nPKCs (PKCθ and PKCδ), IRS/P13, GLUT4 and FFA have also been implicated for the development of insulin resistance and the mechanism proposed is fairly complicated.

Type 2 diabetes is often accompanied by other conditions, including hypertension, low serum high-density-lipoprotein (HDL) cholesterol concentrations, high triglycerides and obesity. This constellation of clinical conditions is referred to as the metabolic syndrome.

Hepatitis C Virus and Insulin Resistance

Infection of hepatitis C virus (HCV) can make people 3-4 times more likely to develop Type 2 diabetes and insulin resistance. Link between HCV infection, insulin resistance and diabetes is based on the finding that there

Diabetes: A Comprehensive Treatise for Patients and Care Givers,
by Shamim I. Ahmad and Khalid Imam. ©2014 Landes Bioscience.

is an increase of ferritin (an intracellular protein that stores iron and releases it in a controlled way) levels in patients with HCV infection and diabetes.

β-Cell Dysfunction and Exhaustion

The second and equally important pathogenic factor is the decline in number of β cell in pancreas and its dysfunction found in 50% of diabetic patients at the time of diagnosis. As said above, in response to rising glucose levels, insulin is normally secreted in two phases: phase 1, for 0–10 minutes and phase 2, for 10–120 minutes, which continues as long as necessary to maintain euglycemia. Once the fasting glucose levels reach 115–120 mg/dL the first-phase insulin secretion stops. By the time impaired glucose tolerance develops with 2-hour post challenge glucose levels reach 141–199 mg/dL, β cell function is already reduced by 60–70%.

Based on the theory under normal circumstances insulin resistance increases the secretory function of the β cell. Based on this theory, it is assumed that over a prolonged period of time, the increasing demand associated with increasing resistance will result in "exhaustion" of the β cell so that it will ultimately fail. Failure to adequate adaptation to insulin resistance may be due to a genetically programmed β cell abnormality associated with an inability of the normal β cells to adapt to insulin resistance and increased secretory demand thus uncovering a defect in β cell function.

Reduction in β-Cell Mass

Significant reduction or complete depletion of β cell mass normally occurs in Type 1 diabetes and results in absolute insulin deficiency. The normal function of immune system in Type 1 diabetic patients becomes impaired and attacks the insulin-producing β cells present in pancreas. This results in the β cells being destroyed and the body no longer can produce enough insulin to regulate blood-glucose levels. That is the reason why the Type 1 diabetic patients solely dependent on insulin therapy to maintain euglycemia.

Incretins

Incretins are a group of gastrointestinal hormones playing a number of important roles in our digestive systems. It can also affect the amount of food we consume. Hence they are indirectly involved in: (a) increasing the amount of insulin production after food is taken even before blood glucose levels are increased, (b) maintaining the sugar level in the blood, (c) inhibiting glucagon release from the α cells of the islets of Langerhans during hyperglycemia, (d) slowing down gastric emptying and motility leading to less postprandial glucose excursion, and finally (e) decreasing food intake due to increased level of satiety. Two main factors responsible for these activities are glucagon-like peptide-1 (GLP-1) and gastric inhibitory peptide or GIP. Both GLP-1 and GIP are subsequently inactivated by the enzyme dipeptidyl peptidase-4.

7

CHAPTER 8

Biochemical Factors Playing Roles in Diabetes

Adipose Tissues and Adipokines

Human body contains a kind of tissues known as adipose tissues made up of fat cells which have been recognized as a major endocrine organ, producing a number of hormones. Adipose tissues are derived from lipoblasts and are composed of adipocytes. Although adipose tissues' main role is to store energy in the form of lipids, and also modulates glucose homeostasis, it is now well established that the adipose tissue mass is increased in obese people and this plays significant roles in obesity related diseases including diabetes and heart disease.

Two types of adipose tissue have been discovered. Their formation appears to be controlled in part by the adipose gene(s). The adipose tissue in the body consists predominantly of the white fat that contributes to energy storage and wide ranging metabolic regulations.

White Adipose Tissue

White adipose tissue is now recognized to be an active participant in energy homeostasis and physiological functions. Adipose tissue retinol-binding protein 4 (RBP4) and pro-inflammatory cytokines such as tumor necrosis factor α (TNF-α) and interleukin 6 (IL-6) have regulatory roles in the development of insulin resistance and metabolic complications associated with obesity to promote Type 2 diabetes. Other adipokines include chemerin, plasminogen activator inhibitor 1 (PAI-1).

Brown Adipose Tissue

Brown adipose tissues (BAT) are present in children as well as in adults and have a unique chemical structure and a specific metabolic role. It has the ability to dissipate energy by producing heat rather than storing energy as triglycerides and hence it has been a good target for increasing energy expenditure and thereby overcome obesity.

A major difference between white adipose tissue and brown adipose tissue is that the former contain one large globule of triglycerides which

Diabetes: A Comprehensive Treatise for Patients and Care Givers,
by Shamim I. Ahmad and Khalid Imam. ©2014 Landes Bioscience.

displaces the cell's nuclei and other cellular organelles excentrically to the cell's periphery. Brown adipose tissue, on the other hand, contains numerous smaller droplets of triglycerides, much higher number of mitochondria and a specific mitochondrial uncoupling protein-2 or thermogenin. Data, both, from animal and human studies suggest that brown adipose tissues and mitochondrial uncoupling protein-2 can be targeted for interventions to prevent and treat obesity.

Leptin

Leptin is largely produced and secreted by white adipose tissue and constitute the most important signaling component for the regulation of food intake and energy homeostasis. It signals the hypothalamus about the quantity of stored fat. It decreases food intake and increases energy expenditure, thereby indirectly promoting insulin sensitivity.

Adiponectin

Adiponectin is a collagen-like protein exclusively produced and secreted by adipose tissue. This protein improves insulin-sensitizing effects in whole body, as it enhances inhibition of hepatic glucose output as well as glucose uptake and utilization in fat and muscle. In addition to its insulin-sensitizing effects, adiponectin may alter glucose metabolism through stimulation of pancreatic insulin secretion.

In humans, adiponectin levels were shown to be positively correlated with markers of insulin sensitivity in frequently sampled intravenous glucose tolerance tests and clamp studies, yet negatively correlated with adiposity, insulin resistance, Type 2 diabetes and metabolic syndromes. Other studies indicated that lower adiponectin levels were also associated with a higher incidence of Type 2 diabetes.

Ghrelin

Ghrelin is a peptide hormone which is produced mainly by gastric and pancreatic cells. The level of ghrelin is increased by hunger and decreases after meals. It acts as the counterpart of the hormone leptin. Thus ghrelin is one of the circulating peptides, which stimulates appetite and regulates energy balance, and thus its gene is one of the candidates to be targeted for obesity and Type 2 diabetes. Recent studies show that insulin inhibits ghrelin secretion in healthy normal weight and overweight persons, and both oral and intravenous glucose loads are also shown to regulate ghrelin secretion in humans. Although studies on ghrelin is still in progress these results points the important roles of this peptide in hunger, obesity and Type 2 diabetes.

Retinol Binding Protein 4

Secreted by adipocytes, retinol binding protein 4 is an adipokine belonging to lipocalin family and acts as carrier of retinol (Vitamin A alcohol) in the blood. This protein is also involved in insulin signaling pathway, insulin resistance and the metabolic complications of obesity, although the data are currently conflicting on this issue. It can also act as a signal to other cells when there is decrease in plasma glucose levels. More studies are needed to understand the role of this protein in diabetes.

Tumor Necrosis Factor *α*

Tumor necrosis factor-alpha (TNF*α*) is a pro-inflammatory cytokine produced by adipocytes and mostly secreted by infiltrating macrophages in the stromal vascular fraction. TNF*α* levels are known to increase in obesity and may play a major role in the impairment of insulin action via disturbing insulin signaling. At molecular level, TNF*α* increases serine phosphorylation of IRS-1, down-regulates GLUT4 expression and inhibits insulin-induced glucose uptake in healthy subjects thereby contributing to insulin resistance.

8

Plasminogen Activator Inhibitor 1

Plasminogen activator inhibitor 1 (PAI-1), an inhibitor of fibrinolysis, is another protein related to adipocytes. It is secreted from endothelial cells, mononuclear cells, hepatocytes and fibroblasts and has been associated with an increased risk for cardiovascular disease. Research has found that increased expression of PAI-1 in the heart is profibrotic and also accompanies insulin resistance in Type 2 diabetes as a factor contributing to the high incidence of heart failure after myocardial infarction in people with diabetes.

Obestatin

Obestatin is a small peptide hormone, released primarily from the gastrointestinal tract, and also from the spleen, mammary gland, breast milk and plasma. In contrast to ghrelin, which causes increased hunger or hyperphagia resulting in obesity, obestatin decreases the desire for food intake and hence counteracts obesity. Thus this is an important hormone playing roles not only in general physiology but also in obesity, diabetes and in people suffering from psychogenic eating disorders such as bulimia. Circulating obestatin concentrations are decreased in individuals with diabetes and impaired glucose tolerance compared with normal glucose tolerance. Full biochemical studies of circulating obestatin have yet to be carried out.

Glucagon

Glucagon is an important hormone secreted by the pancreas and its main role is to increase blood sugar levels. When blood sugar level goes down below the required levels, pancreas releases glucagon causing liver to convert stored glycogen into glucose, a process known as gluconeogenesis, thereby increasing plasma glucose level. When the blood glucose levels are increased more than required levels the release of insulin is stimulated and glucagon inhibited from pancreas. Insulin allows glucose to be taken up and be used by insulin-dependent tissues. Thus, glucagon and insulin are part of a feedback system that keeps blood glucose levels at the stable levels.

CHAPTER 9

Environmental Factors Inducing Type 1 and Type 2 Diabetes

While genetic studies of Type 1 diabetes has branded it as an autoimmune disease involving mostly CD4 and CD8 T cells destroying pancreatic β cells, it is anticipated that the genetic alone does not account for the variance in incidence and severity of Type 1 diabetes. Significant evidence exists in support of the critical role of environmental factors in its development. Studies indicate that only 13–33% of monozygotic twins are pair wise concordant for the disease, which suggests that differential exposure to putative environmental factors may influence the progression of the disease. Several exogenous factors have been studied, and the most promising of these factors are summarized below.

Viral Infection

Studies in humans and animal models clearly suggest that viral infections can modulate autoimmune responses. Evidence from epidemiological studies and animal models demonstrate a dual role of viruses as inducers and accelerators of disease, as well as a protective role supporting the so-called 'hygiene hypothesis.' Several viral infections such as Coxsackie virus B, Varicella-zoster, certain enteroviruses, Rotavirus, Cytomegalovirus, Rubella and Mumps have been suggest as putative triggering agents.

In particular, studies into enteroviruses, that infect β cells, support the hypothesis that certain viral infections can trigger β cell autoimmunity. Enteroviruses have been shown to target and destroy islet β cells in vitro, and anti-enterovirus antibodies (as well as enterovirus RNA) are more frequently found in newly-diagnosed cases of Type 1 diabetes as compared to healthy individuals. Further epidemiological and experimental evidence supports the hypothesis that enteroviruses play a role in the development of Type 1 diabetes by altering β cell function and viability. Nevertheless, a better understanding of the cellular and molecular mechanisms underlying the interaction between virus and host is required to better understand its role in diabetes development.

Diabetes: A Comprehensive Treatise for Patients and Care Givers,
by Shamim I. Ahmad and Khalid Imam. ©2014 Landes Bioscience.

Due to large-scale vaccination programs, viruses such as Mumps and Rubella have been eradicated in many countries with rising incidences of autoimmune diseases such as Type 1 diabetes. The "hygiene hypothesis" attempts to explain why the incidence of diseases such as Type 1 diabetes is higher in industrialized countries with better sanitation and lower rates of viral infections. The theory is that frequent exposure to certain viruses can ultimately protect children from Type 1 diabetes by potentiating their immune systems, allowing a stronger and more rapid immune response upon acute infection.

While evidence implicates a wide variety of viruses, they are thought to be one of many potential factors that influence disease progression. Further complicating the picture is the 'hygiene hypothesis' which shows that still more exogenous factors can influence the relative importance of viral infections on a case-by-case basis. The sensitive balance of the immune system is continuously affected by numerous environmental factors, and the over-arching challenge is to determine what effect the sum of all environmental factors that we encounter in our lifetime has on the etiology of autoimmune diseases such as Type 1 diabetes.

Cow's Milk

As it is the only β cell-specific auto-antigen in postnatal life, insulin auto-antigens are often the first to be detected in young children with preclinical diabetes. The frequency of insulin auto-antigens is significantly higher in young children compared to older children and adults at the time of diagnosis of diabetes. Cow's milk contains bovine insulin and is often an infant's most common source of foreign complex proteins in most developing and also developed countries and its potential role as a trigger for Type 1 diabetes has been studied. Bovine insulin is very similar to human insulin, differing only by three amino acids. It is possible that in young patients exhibiting β cell autoimmunity, the initial immune response to bovine insulin may change in nature to become a response to endogenous insulin. Nevertheless, further studies are needed to make a definite conclusion on the role of cow's milk in the pathogenesis of Type 1 diabetes.

Toxins

Extrinsic factors that damage β cell function directly trigger the autoimmune reaction, including but are not limited to toxic chemical agents such as nitrophenyl urea in rat poison, and cyanide from spoiled food.

Bacterial, Fungal and Parasitic Infections

The role of microbial infections in accelerating the progression of Type 1 diabetes remains unclear. However, infections have the potentials to stimulate autoimmunity. Proposed mechanisms by which microbial infections may contribute to disease pathogenesis are (a) molecular mimicry of self antigens, (b) bystander activation of lymphocytes, and (c) alteration of cytokine patterns near the islet β cells.

White Rice Consumption

In a recent study published in British Medical Journal it has been claimed that high intake of white rice is associated with a significantly increased risk of developing Type 2 diabetes. The studies were carried out in two groups: the heavily rice consuming groups, China and Japan and non-rice consuming groups, USA and Australia. Exact association between white rice consumption and Type 2 diabetes is yet to be discovered.

Vitamin D

In humans and in animal models of diabetes, vitamin D deficiency has been shown to impair insulin synthesis and secretion. Further studies are required to confirm the importance of vitamin D in diabetes

CHAPTER 10

Signs and Symptoms of Diabetes

Although many signs and symptoms have been associated with diabetes, it is not necessary that all of them are present in each of the diabetic patients. Hence it is advisable that if a person is experiencing even some of those symptoms they should immediately consult their doctors. It should also be noted that a large number of diabetic patients are asymptomatic and diagnosed incidentally. This is the reason why Type 2 diabetes normally diagnosed, on the average, 10 years after the start of the disease and even then it is usually picked up during routine medical examination. This is the main reason that in any population at any one time there are on an average only about 50% of the diabetic sufferers become registered diabetics.

Common Symptoms of Diabetes

Polydipsia

Polydipsia or excessive thirst is one of the common features of diabetes. It may very well be ignored by the patients specially those working in the open fields and those living in tropical countries where the summer temperature can reach as high as 50–55°C.

Polyurea

Polyurea or passing urine more frequently is another common symptom of diabetes. It is a consequence of osmotic diuresis resulting in a loss of glucose as well as free water and electrolytes in the urine. Some patients may pass frequent urination especially at night, a phenomenon called as nocturia.

Polyphagia

Polyphagia means excessive hunger or appetite. Diabetic patients often feel hungrier. When plasma glucose level remains elevated, glucose from the blood cannot enter into the cells because of lack of insulin or resistance. Therefore, body cannot convert food in to energy and causes increased hunger.

Diabetes: A Comprehensive Treatise for Patients and Care Givers,
by Shamim I. Ahmad and Khalid Imam. ©2014 Landes Bioscience.

Weight Loss

Weight loss is also a common feature of diabetes. It is usually due to depletion of water which constitutes more than 60% of body weight.

Erectile Dysfunction

It is an inability to achieve and maintain an erection sufficient for sexual intercourse is a distressing and common symptom in diabetes and reported almost 50% in men aged 55 to 59 years. In addition to increasing age, peripheral and autonomic neuropathy also contributes significantly to this problem.

Delayed Wound Healing

Diabetic patients experience their wound healing going slowly; this is due to impaired immune system.

Recognized Features

Obesity

Many patients with Type 2 diabetes are overweight or obese. Even those who are not significantly obese often have raised waist circumference.

Acanthosis Nigricans

Some patients may have acanthosis nigricans, which is associated with significant insulin resistance; the skin in the axilla, groin and back of neck becomes hyper pigmented and hyperkeratotic.

Hyperglycemic Hyperosmolar State/Diabetic Ketoacidosis

In these cases, patients are profoundly dehydrated, hypotensive, lethargic or comatose but without ketoacidosis. Mortality in this condition is as high as 15%. This is more commonly in Type 2 diabetes while diabetic ketoacidosis is more common in Type 1 diabetes.

CHAPTER 11

Diagnostic Criteria and Tests for Diabetes

Diabetes mellitus can be diagnosed by doing following blood tests:
1. Fasting plasma glucose
2. Random plasma glucose
3. Oral glucose tolerance test
4. HbA1$_C$ test

Fasting Plasma Glucose

This is determined in the morning before any food or drink taken (at least 8 hour or no more than 14 hours). A measurement of greater than 126 mg/dL on more than one occasion is diagnostic of diabetes. False positive may result if the patient is taking corticosteroids, contraceptive pills, diuretics or excess thyroxin.

Random Plasma Glucose

The level of ≥200 mg/dL at any time during the day regardless of the time of the last meal on more than one occasion or random glucose of more than 200 mg/L with the classic symptoms of excessive urination, thirst and unexplained weight loss.

Oral Glucose Tolerance Test

If the plasma glucose test is not conclusive then oral glucose tolerance test is carried out. For this (a) patients should consume at least 150–200 g of carbohydrate for 3 consecutive days before the procedure (b) they should also observe an overnight fasting for at least 8 hours. (c) no tea, coffee, cigarette smoking or exercise allowed during the test (d) 75 g of glucose dissolve in 300 ml water should be ingested (e) plasma glucose should be checked in fasting and 2 hours after drinking the glucose solution. Cut of values for fasting and random are the same as above.

Glucose concentrations are expressed as milligrams per deciliter (mg/dL or mg/100 mL) in the United States, Japan, Spain, France, Belgium, Egypt, Saudi Arabia and Colombia, while millimoles per liter (mmol/L) are the units used in most of the rest of the world. Glucose concentrations

Diabetes: A Comprehensive Treatise for Patients and Care Givers,
by Shamim I. Ahmad and Khalid Imam. ©2014 Landes Bioscience.

expressed as mg/dL can be converted to mmol/L dividing by 18 (the molar mass of glucose). For example, a glucose concentration of 90 mg/dL is 5.0 mmol/L (90 ÷ 18= 5). Multiply by 18 if converting mmol to mg/dL.

HbA1$_C$ Test

Another important and reliable method of measuring blood glucose level is to determine the amount of glycosylated hemoglobin. Since 2009 HbA1$_C$ measurement is strongly recommended for the diagnosis of diabetes. HbA1$_C$ ≥6.5% is diagnostic of diabetes.

The advantage of using the HbA1$_C$ to diagnose diabetes is that there is no need to fast; it has lower intraindividual variability than the fasting glucose and the oral glucose tolerance test.

The reason for the importance of this test is that glucose levels in human blood vary throughout the day depending upon the food consumed and the treatment received for diabetes. Hence, instead of measuring the glucose levels in blood many times a day and then determine the average value, the HbA1$_C$ test has been developed to measure the level of glucose in the blood for the previous 3 months and represent their average reading for this period. This is because the red blood cells survive in the system for around 8–12 weeks before they are replaced by fresh cells synthesized in the body. HbA1$_C$ measures the levels of glycated hemoglobin (HbA1$_C$ or hemoglobin A1$_C$) in the blood. The formation of hemoglobin A1$_C$ is due to attachment of glucose molecules in hemoglobin in red blood cells. The more the glucose in the blood the higher is the glycated hemoglobin or HbA1$_C$. Hence the test is the best and most effective measure of accurate treatment (antidiabetic tablets and/or insulin, exercise etc) and in controlling long term diabetes complications. Interestingly, short term or hourly elevations in blood sugar levels do not seem to affect the total HbA1$_C$ value.

In non-diabetic person HbA1$_C$ comprises 4.5–6% of total hemoglobin A1$_C$. The hemoglobin A1$_C$ fraction is abnormally elevated in diabetic patients with uncontrolled diabetes. A good range of HbA1$_C$ in diabetic patient is below 7.0

Concerns should be raised for diabetic complications if the HbA1$_C$ levels persistently show more than 7%.

Factors Affecting HbA1$_C$

HbA1$_C$ test is inappropriate in populations with high prevalence of haemoglobinopathies or in conditions with increased red cell turnover. Thus, incorrect high values may be obtained when red cell turnover is slow, resulting in an incorrect number of older red cells. This problem can occur in patients with iron, vitamin B$_{12}$ or folate deficiency anemia. On the other hand, rapid red cell turnover leads to a greater proportion of younger red cells and falsely displaying low HbA1$_C$ values. Examples include patients with haemolysis and those treated for iron, vitamin B$_{12}$, or folate deficiency and patients treated with erythropoietin. Also false test results may be obtained due to vitamins C and E which is possibly due to inhibiting glycation of hemoglobin. Thus care should be taken to avoid these false results. Also, the test should be performed using a National Glycohaemoglobin Standardization Program certified method and standardized to the Diabetes Control and Complications Trial assay.

Global Unification of HbA1$_C$ Value

The laboratory results for HbA1$_C$ are presented either in percentage or in mmol/L. International Federation of Clinical Chemistry has adopted the measurement method in mmol/mol. A table has been drafted for converting the % data into mmol/mol (see Table 11.1).

Table 11.1. HbA1$_C$ Conversion Table

HbA1$_C$ *(mmol/mol)	HbA1$_C$ (%)
31	5
37	5.5
42	6
48	6.5
53	7
59	7.5
64	8
69	8.5
75	9
80	9.5
86	10
91	10.5
97	11
102	11.5
108	12

*HbA1$_C$ Conversion Formula:
HbA1$_C$ mmol/mol = [HbA1$_C$ % - 2.15] × 10.929
HbA1$_C$% = [HbA1$_C$ mmol/mol ÷ 10.929] + 2.15

Glycosuria as a Diagnostic Test for Diabetes

In early period of the diagnosis of diabetes, glycosuria test was considered as only method for detecting diabetes. The test involved a paper strip impregnated with glucose oxidase and a chromogen system, which is sensitive to as little as 0.1% of glucose in urine. A normal renal threshold for glucose as well as reliable bladder emptying is essential for interpretation.

Glucose in urine may appear in certain benign or pathological conditions other than diabetes and it may not appear in urine despite mild hyperglycemia because of a particular renal threshold for glycosuria. Therefore, urinary glucose test is generally not carried out for the diagnostic purpose.

11

CHAPTER 12

Complications of Diabetes Mellitus

In the human body a number of biochemical reactions function in synchrony to bring about and maintain a healthy physiological state. At the core of these processes lies the ability of the man to maintain a constant stable state or homeostasis. An aberration of the homeostatic state leads to the development of an injury or a pathological state in various organs. Diabetes reduces the ability of the patients to regulate the level of glucose in the blood stream resulting in acute and chronic complications.

Acute Complications

Diabetic Ketoacidosis

Diabetic ketoacidosis (DKA) is a term used when the excess sugar in the blood is converted to the acid called ketones. In normal person glucose enters the cells where it can be used as energy but in diabetic patients the lack of glucose entering the cells and its subsequent accumulation may lead to its conversion in ketones.

Ketones being very harmful, the body tries to remove it by excreting them with the urine. If the level of ketones in body keeps on increasing it leads to increased ketones in the blood. Subsequently nausea or vomiting may start and this can lead to more dehydration and less effective to remove ketones. Other effects of ketoacidosis are dry skin, blurred eyesight and breathlessness. Further rise in ketones can lead to bad breath. If untreated, the further rise of ketones in blood can lead the patient to enter in coma and be fatal. Hence it is very important to diagnose DKA as early as possible. Ketoacidosis is a life-threatening condition which needs immediate treatment.

Ketoacidosis usually develops when the blood glucose level is more than 250 mg/dL or 14 mmol/L and it can be detected by a simple urine test, but measuring ketones in blood is more sensitive in diagnosing ketoacidosis. Ketoacidosis is more common in patients with Type 1 diabetes and uncommon in Type 2 diabetes. Interestingly, however, ketoacidosis is more common in Afro-Caribbean and Hispanic obese subjects with Type 2 diabetes. It is also more common in diabetic patients during certain kinds of illness when body's response to illness (such as infection) leads to release of more glucose in blood stream. Hence when a patient falls ill, he should check his blood glucose level more often and keep a watch on the symptoms of ketoacidosis.

Hyperglycemic Hyperosmolar State/Hyperglycemic Non-Ketotic State

In hyperglycemic Hyperosmolar state (HHS), patients are profoundly dehydrated, hypotensive, lethargic or comatose. Mortality in this condition is as high as 15%.

Diagnostic Criteria of DKA/HHS

According to the consensus statement published by the American Diabetes Association, diagnostic features of HHS and DKA may include the following:

HHS
- Plasma glucose level of 600 mg/dL or more
- Effective serum osmolarity of 320 mOsm/kg or more
- Profound dehydration, up to an average of 9L
- Serum pH greater than 7.30
- Bicarbonate concentration greater than 15 mEq/L
- Small ketonuria and absent-to-low ketonemia
- Some alteration in consciousness

DKA
- Plasma glucose >250 mg/dL (13.9 mmol/L)
- Arterial pH <7.3
- Ketonemia and/or ketonuria
- Bicarbonate concentration less than 15 mEq/L

Management of DKA/HHS

In general, treatment of DKA and HHS requires aggressive correction of dehydration, hypovolemia and hyperglycemia, replacement of electrolyte losses and careful search for the precipitating cause(s). A flow sheet is invaluable for recording vital signs, volume and rate of fluid administration, insulin dosage, and urine output and to assess the efficacy of medical therapy. In addition, frequent laboratory monitoring is important to assess response to treatment and to document resolution of hyperglycemia and/or metabolic acidosis. Serial laboratory measurements include glucose and electrolytes and, in patients with DKA, venous pH, bicarbonate, and anion gap values until resolution of hyperglycemia and metabolic acidosis.

Fluid Therapy

The cornerstone of DKA and HHS management is hydration therapy. Patients with DKA and HHS are invariably volume depleted, with an

12

estimated water deficit of: 100 ml/kg of body weight. The initial fluid therapy is directed towards expansion of intravascular volume and restoration of renal perfusion. Isotonic saline (0.9% NaCl) infused at a rate of 500–1,000 mL/h during the first 2 hours then normal saline infusion should be reduced to 250 mL/h or changed to 0.45% saline (250–500 mL/h) depending on the serum sodium concentration and state of hydration. The goal is to replace half of the estimated water deficit over a period of 12–24 hours. Care should be taken to replace intravenous fluids in elderly patients so as to minimize risk of pulmonary edema.

Once the plasma glucose reaches 250 mg/dL in DKA and 300 mg/dL in HHS, replacement fluids should contain 5–10% dextrose to allow continued insulin administration until ketonemia is controlled while avoiding hypoglycemia.

Insulin Therapy

Insulin increases peripheral glucose utilization and decreases hepatic glucose production, thereby lowering blood glucose concentration. In addition, insulin therapy inhibits the release of free fatty acids from adipose tissue and decreases ketogenesis, both of which lead to the reversal of ketogenesis.

In critically ill and mentally obtunded patients, regular insulin given intravenously by continuous infusion is the treatment of choice. Such patients should be admitted to an intensive care unit or to a step down unit where adequate nursing care and quick turnaround of laboratory tests results are available. An initial intravenous bolus of regular insulin of 0.15 unit/kg of body weight, followed by a continuous infusion of regular insulin at a dose of 0.1 unit/kg/h should be administered. When plasma glucose levels reach 250 mg/dL in DKA or 300 mg/dL in HHS, the insulin infusion rate should be reduced to 0.05 unit/kg/h (3–5 units/h), and dextrose (5–10%) should be added to intravenous fluids.

Potassium Replacement

Despite a total body potassium deficit of 3–5 mEq/kg of body weight, most patients with DKA have a serum potassium level at or above the upper limits of normal. These high levels occur because of a shift of potassium from the intracellular to the extracellular space due to acidemia, insulin deficiency, and hypertonicity. Both insulin therapy and correction of acidosis decrease serum potassium levels by stimulating cellular potassium uptake in peripheral tissues. Therefore, to prevent hypokalemia, most patients require intravenous potassium infusion during the course of DKA therapy. The treatment goal is to maintain serum potassium levels within the normal range of 4–5 mEq/L.

Criteria of Resolution of DKA and HHS

DKA

The criteria for resolution of ketoacidosis include:
- Blood glucose <200 mg/dL
- Serum bicarbonate level ≥18 mEq/L
- Venous pH >7.3
- Calculated anion gap ≤12 mEq/L

HHS

The criteria for resolution of HHS include:
- Improvement of mental status
- Blood glucose <300 mg/dL
- Serum osmolality of <320 mOsm/kg

Switching to Subcutaneous Insulin

Patients with moderate to severe DKA should be treated with continuous intravenous insulin until ketoacidosis is resolved. When these levels are reached, subcutaneous insulin therapy can be started. If patients are able to eat, split-dose therapy with both regular (short-acting) and intermediate-acting insulin may be given.

Patients with already diagnosed diabetes may be given insulin at the dosage they were receiving before the onset of DKA. In patients with newly diagnosed diabetes, an initial insulin dose of 0.6 unit/kg/day is usually sufficient to achieve and maintain metabolic control. Two-thirds of this total daily dose should be given in the morning and one-third in the evening as a split-mixed dose.

Diabetes Education

The introduction of diabetes educational programs in most diabetes clinics have contributed to a reduction in the occurrence of DKA and HHS in patients with known diabetes. Such programs teach patients how to avoid these complications by self-testing for urinary ketones and adjusting their insulin regimens on sick days. It is essential to educate patients in the prevention of DKA so that a recurrent episode can be avoided. Central to patient education programs for adults with diabetes is instruction on the self-management process and on how to handle the stress of inter-current illnesses.

Hypoglycemia and Whipple's Triad

Hypoglycemia is the state in which the blood sugar becomes lower than normal. Hence hypoglycemia is the opposite physiological state of hyperglycemia and truly defined by Whipple's Triad:
- Low plasma glucose level
- Symptoms of hypoglycemia
- Correction of symptoms of hypoglycemia on administering glucose

Hypoglycemia is more dangerous than hyperglycemia because our brain activities are dependent on a continuous supply of glucose diffusing from the blood into the interstitial tissue within the central nervous system and into the neurons themselves. If the amount of glucose falls below 2.2 mmol/L or 40 mg/dL, the brain is one of the first organs to be affected. The affect can include reduction of mental efficiency, impairment of action and judgment. If the level of glucose falls further, seizure, coma and death may occur. It is difficult to determine exactly at what point of the blood sugar the hypoglycemia starts because this phenomenon is different for different person, circumstances and for purposes, and has been a matter of controversy.

Certain poisons, alcohol and hormone deficiency have also been known to induce hypoglycemia. Also long term starvation, alterations of metabolism associated with infection and organ failure can also induce hypoglycemia.

If identified in time, hypoglycemia can be successfully treated without causing any damage to the system. The treatment which can raise the blood sugar levels to normal is the ingestion of a suitable source of carbohydrate such as sweet biscuit, bar of chocolate or fruit such as banana or orange which can be taken as food or drink if the person is able to swallow. Alternatively the carbohydrate can be taken from a suitable fruit juice. About 100-120 ml of any suitable fruit juice should be enough to eliminate hypoglycemia within 15-20 minutes. To be noted that fruit juice contains more fructose (a kind of sugar) and this enters in the metabolic system more slowly than glucose. Many other sugar-containing food and drinks exists (such as soft sugary drinks, slices of bread and whole fruits) can also work if given soon after experiencing hypoglycemia. If, on the other hand, hypoglycemia has entered in its advance stage, dextrose or glucose should be preferred sugar to be taken. If hypoglycemia has reached to the level in which the person is unconscious and cannot swallow or eat or drink any thing, in that case patient should be immediately transferred to the hospital for the administration of intravenous glucose.

Hypoglycemic Symptoms

Hypoglycemia can be identified from the symptoms as below.

Autonomic Symptoms

- Sweating
- Palpitations
- Tremulousness
- Anxiety
- Hunger

Neuroglycopenic Symptoms
- Confusion
- Difficulty with concentration
- Irritability
- Hallucinations
- Focal deficit
- Coma (and possible death)

Hypoglycemia in Children

It is important to know the circumstances why hypoglycemia in children can occur. Also when a child has been diagnosed to be suffering from diabetes, in the beginning it may be difficult to recognize when the child is suffering from hypoglycemia. For this reason it is important to have a more frequent blood glucose checking at home.

Hypoglycemia may happen because of the following reasons:
- A delayed or missed meal
- Too much insulin administered
- Inadequate meal
- Heavier physical activities

If a child becomes unconscious due to hypoglycemia, it is important to remember following things:
- Never to give them food or drink from mouth.
- Put the child in his/her side with the head tilted back.
- Keeping a glucagon injection at hand and with full training administer it intramuscularly. Otherwise transfer to the hospital immediately.

Some patients do not experience hypoglycemia at the early phase of diabetes. In them the release of adrenaline, which normally warns of this phenomenon, does not occur until after their brains are significantly deprived of glucose. But when the hypoglycemia attack occurs it can be severe, in that they feel drowsiness or becoming unconscious; this phenomenon is known as 'hypoglycemia unawareness'.

Lack of early sign of hypoglycemia is more common among Type 1 diabetic patients specially those who have long history of diabetes or have been experiencing frequent hypoglycemia. The reason for this is that in an early stage of diabetes the threshold of recognition of symptoms occurs at a lower level of blood glucose but at later stage the threshold of triggering the release of adrenaline falls below the level of blood glucose that triggers the reaction.

Chronic Complications

Microvascular Complications

The mechanism by which poor glycemic control predisposes to microvascular disease is incompletely understood. However, following mechanisms have been postulated.

    a. Accumulation of AGEs in plasma that contribute to microvascular disease

    b. Accumulation of cellular sorbitol, which interferes with cellular metabolism because of a rise in cell osmolality and a decrease in intracellular myoinositol

    c. End-organ response with activation of cytokines, profibrotic elements, vascular growth factors, inflammation, and protein kinase C. Specific end organ responses include mesangial matrix expansion and glomerular hypertension in the kidney, and impairment of retinal blood flow and microthrombus formation in the eye.

Retinopathy, nephropathy, neuropathy and other microvascular complications are described below in full detail.

Retinopathy

Diabetic retinopathy is caused due to damage to the retinal vasculature. It is a common cause of blindness and visual impairment in the working age population.

The first stage of diabetic retinopathy is commonly asymptomatic but with the progression of retinopathy the patients may experience blurred or hazy vision, distortion and visual acuity loss.

Several methods have now been developed including fundoscopy, fluorescein angiography, optical coherence tomography (OCT) and automated screening algorithms. Fundoscopy is most commonly used due to its simplicity, cost and maintenance. Automated screening algorithms are currently being developed to use in conjunction with digital photography. If successfully developed, the use of automated detection systems will allow quick and cost-effective diagnosis of retinopathy at an early stage of the disease allowing timely treatment.

Classification of Retinopatyh

Retinopathy can be classified in to (i) non-proliferative diabetic retinopathy (NPDR) and (ii) proliferative diabetic retinopathy (PDR). It progresses from mild to moderate and severe NPDR and finally to PDR, characterized by the abnormal growth of new blood vessels (iii) diabetic maculopathy or diabetic macular oedema (DMO), caused by leakage of fluid from blood vessels and retinal thickening can develop at any stage of retinopathy and (iv) advanced diabetic eye disease (ADED).

Non-proliferative diabetic retinopathy. The earliest clinical feature of NPDR is the presence of microaneurysm. Microaneurysms are small swellings which form on the side of tiny blood vessels. These swellings sometime break and allow blood to leak into nearby tissue. People with diabetes are more prone of microaneurysm and the first sign of eye damage caused by many years of high blood glucose levels. Good glycemic control may reverse microaneurysms .

Proliferative diabetic retinopathy is the growth of new blood vessels in the retina and it occurs at a later stage and 2.5 times more common in patients with Type 1 than Type 2 diabetes.

Diabetic maculopathy is the most common cause of visual loss in Type 2 and to some extent in Type 1 diabetes. It is characterized by the presence of hard exudates, due to increased vascular leakage, and is a condition in which fluid accumulates at the macula causing visual impairment.

Advanced diabetic eye disease (ADED). Uncontrolled PDR eventually leads to ADED which eventually leads to complete blindness and is characterized by vitreous hemorrhages, tractional retinal detachments, and neovascular glaucoma and rubeosis iridis.

Risk Factors of Retinopathy

The major risk factor for diabetic retinopathy is poor glycemic control, high blood pressure, longer duration of diabetes and dyslipidemia. Other risk factors include, smoking, ethnicity, cataract extraction, puberty and pregnancy.

Treatment of Retinopathy

A) Glucose control is the most important measure to control diabetic retinopathy. A drop in $HbA1_C$ levels by 1% lowers the risk of diabetic retinopathy by 30-40% and the effect appears to be long lasting.

B) Blood pressure control is another measure to reduce the risk of diabetic retinopathy because hypertension exacerbates diabetic retinopathy through increased blood flow and mechanical damage of vascular endothelial cells. An increase of 10 mm Hg in systolic blood pressure is associated with a 10% risk of early diabetic retinopathy and 15% risk of proliferative diabetic retinopathy. Tight blood pressure control has been demonstrated to reduce the risk of diabetic retinopathy progression by a third, visual loss by half and the need for laser treatment by a third in people with Type 2 diabetes.

Once sight-threatening diabetic retinopathy has been detected, the treatment options are limited. Conventional laser photocoagulation and vitrectomy remain the best option for preventing sight-threatening retinopathy.

Intravitreal Anti-VEGF, bevacizumab has been shown to be effective in diabetic retinopathy. C-peptide, though not commercially available, has shown promising results in the treatment of diabetes complications related to vascular degeneration.

Screening of Retinopathy

- All Type 2 diabetic patients should have a dilated fundoscopic eye examination at the time of diagnosis and annually.
- Type 1 diabetics should have eye examination 3-5 years after diagnosis and annually thereafter. Examination is required more frequently as soon as any sign of retinopathy appears.
- Women with preexisting diabetes who are planning a pregnancy or who have become pregnant should have a comprehensive eye examination starting from the first trimester and should be counseled on the risk of development and progression of diabetic retinopathy.

Nephropathy

Kidney is a vital organ and plays the following important roles:

- Filter and clean the blood by removing all the waste products and extra fluid through urine.
- Regulates the amount of fluid and various salts in the body—thus helping the control of blood pressure.
- Release hormone, erythropoietin, that act on bone marrow for the production of red blood cells.

Diabetic nephropathy is the leading cause of end stage renal disease worldwide. It is defined by proteinuria > 500 mg in 24 hours and preceded by microalbuminuria which is defined as albumin excretion of 30-299 mg/24 hours. Without intervention, diabetic patients with microalbuminuria typically progress to overt nephropathy. As many as 7% of patients with Type 2 diabetes may already have microalbuminuria at the time they are diagnosed with diabetes. In the European Diabetes Prospective Complications Study, the cumulative incidence of microalbuminuria in patients with Type 1 diabetes was 12% during a period of 7 years. Approximately 15-20% of Type 2 diabetic patients and 20-40% of Type 1 patients are expected to develop diabetic nephropathy and end stage renal disease.

Occurrence of nephropathy is a slow process and takes time and may be asymptomatic in the beginning. Bilateral ankle edema is one of the features of nephropathy but usually seen in advanced stage of nephropathy.

Abnormal Protein Excretion

Normal protein excretion is less tha 150 mg/day. Increased urinary protein excretion is the earliest clinical manifestation of nephropathy. However, when assessing protein excretion, the urine dipstick is a relatively

Table 12.1. Abnormalities in Albumin Excretion

Category	Spot Collection microgram/mg Creatinine
Normal	<30
Microalbuminuria	>30–299
Macroalbuminuria	≥300

insensitive marker for initial increases in protein excretion, not becoming positive until protein excretion exceeds 300 to 500 mg/day. Although 24-hour urine collection used to be the initial gold standard for the detection of albuminuria, it has been suggested that screening can be more simply achieved by a timed urine collection or measurement of the urine albumin concentration using the early morning specimen. See Table 12.1 for abnormal urinary albumin excretion.

Measurement of the urine albumin/creatinine ratio in random urinary samples may be preferred as the screening strategy for microalbuminuria in all diabetic patients. If not found at the initial screening, the ratio should be measured on yearly basis in diabetic patients. Variability in urinary albumin excretion can exist hence 2-3 specimens should be collected within 3 to 6 month period before considering a patient to have microalbuminuria. Following conditions may give rise to false positive result:

- Exercise within 24 hours
- Infection
- Fever
- Marked hyperglycemia
- Marked hypertension

The onset of microalbuminuria in Type 1 diabetes is typically between 10 and 15 years after the onset of the disease. Epidemiologic studies have shown that 20 to 40% of the patients with diabetes develop nephropathy, irrespective of glycemic control. It is important to keep control of blood pressure in the range of 130/80 mm Hg or less and HbA1$_c$ below 7%.

Treatment of Nephropathy

The rate of renal function decline, after the development of nephropathy, is highly variable between patients and is influenced by additional factors, including glycemic level, blood pressure, and albuminuria, smoking and dyslipidemia etc.

Once irrereversble renal damage occurs, there is no cure for it but measures can be taken to minimize further damage. The treatment may

12

also help in reducing cardiovascular mortality. Angiotensin Converting Enzyme Inhibitors (ACEIs) and Angiotensin Receptor Blockers (ARBs) are commonly used to prevent the albumin excretion and progression to kidney failure.

In diabetic nephropathy following measures are also required to minimize complication.

Good Glycemic Control

Glycemic control can partially reverse the glomerular hypertrophy and hyperfiltration that are thought to be important pathogenic pathways for nephropathy and decreases the incidence of newly-onset microalbuminuria.

Blood Pressure Control

Tight blood pressure control is important for preventing progression of nephropathy and other complications in patients with diabetes. Hence as soon as nephropathy detected, the recommended thresholds to initiate treatment to bring blood pressure to 130/80 and 125/75 mm Hg for patients with diabetes and diabetes associated nephropathy, respectively. Studies have shown that ACE inhibitors and ARBs are better drugs in reducing disease progression in diabetic patients with nephropathy. ACE inhibitor therapy significantly impedes progression to clinical proteinuria and prevents the increase in albumin excretion rate in non-hypertensive patients with Type 1 diabetes and persistent microalbuminuria.

As salt has been associated with high blood pressure, it is advisable that nephropathy patients also ought to be on low salt. Evidence also exists that a low sodium diet potentiates the antihypertensive and antiproteinuric effects of antihypertensive agents in diabetes, thus, ensuring that patients who receive ACE inhibitors or ARBs treatment should be on a low sodium diet (<90 mEq/day) and taking an appropriate diuretic should increase the likelihood of achieving the antiproteinuric efficacy of antihypertensive agents.

Control of Dyslipidemia

Hyperlipidaemia is common in diabetic patients and lipid-lowering drug such as statins show renoprotection in a variety of proteinuric glomerular diseases. The progression of nephropathy may be significantly affected by treatment of dyslipidemia.

Low Protein Diet

It is still not very clear how much restriction in protein intake has how much affect on diabetic renal failure, as there is sporadic data that some dietary restriction of protein (0.6 g/kg per day) retarded the progression of renal failure in patients with Type 1 diabetes.

Hemodialysis

In hemodialysis, the patient's blood is pumped through the blood compartment of a dialyzer, exposing it to a partially permeable membrane. The dialyzer is composed of thousands of tiny synthetic hollow fibers. The fiber wall acts as the semi-permeable membrane. Blood flows through the fibers, dialysis solution flows around the outside of the fibers, and water and wastes move between these two solutions. The cleansed blood is then returned via the circuit back to the body. Ultrafiltration occurs by increasing the hydrostatic pressure across the dialyzer membrane. This usually is done by applying a negative pressure to the dialysate compartment of the dialyzer. This pressure gradient causes water and dissolved solutes to move from blood to dialysate, and allows the removal of several liters of excess fluid during a typical 3-5 hour treatment.

Indications for Dialysis

The decision to initiate dialysis or hemofiltration in patients with renal failure depends on several factors. These can be divided into acute or chronic indications.

Acute indications for dialysis
1. Severe Metabolic Acidosis
2. Severe hyperkalemia
3. Drug intoxication
4. Volume overload or pulmonary edema
5. Uremia complications, such as pericarditis, encephalopathy, or gastrointestinal bleeding

Chronic indications for dialysis
1. End stage renal disease
2. Difficulty in medically controlling fluid overload, serum potassium, and/or serum phosphorous

Kidney Transplantation

Kidney transplantation is the organ transplant of a kidney into a patient with end-stage renal disease. Kidney transplantation is typically classified as cadaveric and living-donor transplantation. Living-donor renal transplants are further characterized as genetically related (living-related) or non-related (living-unrelated) transplants, depending on whether a biological relationship exists between the donor and recipient.

Indications

The indication for kidney transplantation is end-stage renal disease (ESRD), regardless of the primary cause. Common diseases leading to

ESRD include malignant hypertension, glomerulonephritis, diabetes mellitus; genetic causes include polycystic kidney disease, a number of inborn errors of metabolism, and autoimmune conditions such as Systemic lupus Erythometosis. Diabetes is the most common cause of kidney transplantation, accounting for approximately 25% of those in the US. The majority of renal transplant recipients are on dialysis at the time of transplantation. However, individuals with chronic renal failure who have a living donor available may undergo preemptive transplantation before dialysis is needed.

Contraindication

Severe vital organ dysfunctions are contraindication to renal transplantation, such as:

- Severe left ventricular dysfunction
- Severe Pulmonary insufficiency
- Hepatic disease
- Concurrent tobacco use and morbid obesity are also among the indicators putting a patient at a higher risk for surgical complications.

Screening for Nephropathy

Perform an annual test to assess urine:albumin excretion and serum creatinine in Type 1 diabetic patients with diabetes duration of 5 years or more and in all Type 2 diabetic patients at the time of diagnosis and annually thereafter.

Neuropathy

Diabetic neuropathies are various neurological disorders caused by prolonged exposure to hyperglycemia. Most disorders are caused due to injuries to both micro and macro blood vessels supplying blood to all different organs including nerves. Amongst other complications which can be diagnosed fairly early and easily in diabetic neuropathies are more complex and difficult to be diagnosed. Further complexity arises because it is not a single disease affecting any specific organ but diabetic neuropathies can affect almost the whole body including brain. Macro and Microvasculatures which are present in the form of net-work all over the body are the targets for the neuropathies. Diabetic neuropathy can affect all peripheral nerves including sensory, motor neurons and the autonomic nerves.

Conditions associated with diabetic neuropathy include third nerve palsy; mononeuropathy; mononeuropathy multiplex; diabetic amyotrophy; a painful polyneuropathy; autonomic neuropathy; and thoracoabdominal neuropathy.

Peripheral Neuropathy

In this neuropathy, at initial stage, decreased sensation and loss of reflexes occurs in the toes of one or both feet and then extends upward. This class of neuropathy is also painful and can give burning feeling, pricking sensation, achy or dull and pins and needles sensation. As a result of loss of sensation this neuropathy can meet with a number of side effects, for example any small injury to feet can go unnoticed for some time up to its infection and ulceration and even end up in leg amputation. Other types of side effects include Charcot's joint, contractures of the toes and hammer toes (see below). It is also important to note that the pain in a leg or foot may also occur due to vitamin B_{12} deficiency, uremia, alcoholism and drug side effects which should be discriminated from the pain caused by diabetic neuropathy.

12

Autonomic Neuropathy

This kind of neuropathy involves the nerves serving to involuntary organs such as heart, lungs, blood vessels, adipose tissue, sweat glands, gastrointestinal system and genitourinary system, and can affect any of these organs. Postural hypotension, drop in blood pressure while standing up is also an important feature of diabetic autonomic neuropathy. This can also give rise to cardiac rhythm abnormality.

Autonomic neuropathy also affects gastrointestinal tract leading to gastroparesis (see below), nausea, bloating, and diarrhea. As a result of impairment of gastrointestinal system, it is possible that the orally taken antidiabetic medicine may not be absorbed efficiently hence their effects may not be as pronounced as should be. Also for these reasons the unwanted bacterial population may increase leading to bloating, excessive gases and diarrhea. Also urinary symptoms may be affected in this neuropathy leading to incontinence of urine, urgency, and retention which may lead to frequent urinary tract infections.

Diabetic Gastroparesis

It is a chronic disorder in which emptying of stomach takes significantly longer time than normal. The disease is connected with damage to the vagus nerve which controls the movement of food all the way from stomach into intestines. Hyperglycemia causes damage to blood vessels carrying oxygen to nerves and resulting in malfunctioning of vagus nerve.

The sign and symptoms of gastroparesis are variable and can be from mild to severe. These include nausea, vomiting, early feeling of satiety during eating, weight loss, abdominal bloating, heart burn and loss of hunger.

There is no satisfactory treatment of gastroparesis but diabetic patients can control it to some extent by a better glycemic control, changing their diet and eating habits and taking prokinetic medicines.

Cranial Neuropathy

Cranial neuropathy commonly affects eyes in which cranial nerves are being damaged. Specifically abducent and oculomotor nerve which controls certain eye muscles are affected and hence eye movement becomes impaired. Also pupil and eyelid can be affected due to cranial neuropathy. Other manifestations of cranial neuropathy are mononeuropathies of the thoracic or lumbar spinal nerves which can give rise to pain that can falsely be experienced as myocardial infarction, cholecystitis or appendicitis.

Treatment for Neuropathy

Unfortunately like all other diabetic complications there is no cure for diabetic neuropathy and only treatments available are to alleviate the symptoms. Tight glucose control, on the other hand, is the best control measure. Besides classical analgesics there are a number of drugs including tricyclic antidepressants, serotonin reuptake inhibitors and antiepileptic drugs which may be used in diabetic neuropathy. A combination of gabapentin and nortryptaline can also be tried. Duloxetine and pregabalin are two new drugs being approved by FDA and may be used. Some other drugs such as imipramine, amitriptyline, desipramine and nortriptyline belonging to tricyclic antidepressants group may be taken with great care due to their cardiac toxicity especially at high doses.

Screening for Diabetic Neuropathy

All patients should be screened for distal symmetric polyneuropathy (DPN) at diagnosis and at least annually thereafter, using simple clinical tests such as tuning fork and monofilament test.

Screening for signs and symptoms of autonomic neuropathy should be instituted at diagnosis of Type 2 diabetes and 5 years after the diagnosis of Type 1 diabetes. Special testing is rarely needed and may not affect management or outcomes.

Macrovascular Complications

Ischemic Heart Disease

Approximately 60 to 70% of patients with diabetes die of cardiovascular diseases and as mentioned above there is a strong link between diabetes and the heart disease. It is especially worrying because about 50% of the diabetic patients remain undiagnosed for about 10 years and during this period, due to little or no control, diabetic patients carry a significant risk to suffer a fatal heart attack. This is the reason why diabetes is considered a coronary heart disease risk equivalent condition. To understand the mechanism of atherosclerosis it is important to understand the role of hyperglycemia in affecting various biological components leading to this kind of complication.

Atherosclerosis in Diabetes

As shown earlier, hyperglycemia is a common problem in every kind of diabetes in which the glucose molecules react with various biological molecules. The reaction usually occurs with the amino groups of amino acids in proteins to form a Schiff's base and then undergo a rearrangement to form a bound ketoamine. For example glucose can react with the amino group of lysine and the product is deoxyfructose-lysyl-protein. This non-enzymatic modification of proteins by glucose is known as glycation. A notable glycated protein found in diabetes is glycated hemoglobin or HbA1$_C$, which is monitored as a measure of glycemic control (see above). Further reactions of modified protein lead to advanced glycation end products (AGEs). These AGEs react with a specific receptor (RAGE) at the vascular endothelium, increasing the production of superoxide anion and other oxidative products which accelerate atherogenesis.

Dyslipidemia Causing Increasing Ischemic Heart

Following atherogenic cholesterols may significantly increase the risk of cardiovascular disease and stroke.

Low density lipoprotein (LDL) cholesterol. This is the most atherogenic cholesterol which can build up inside the blood vessels, leading to hardening and narrowing of arteries carrying blood from the heart to the rest of the body. If the narrowing increased and blood circulation becomes blocked it leads to ischemia and if untreated can be fatal.

Triglycerides are another type of blood fat that can raise the risk of heart disease when the levels are high. Triglycerides mainly come from of vegetable oil and animal fats. High levels of triglycerides in the bloodstream have been associated with atherosclerosis, hence the risk of heart disease and stroke. However, a strong inverse relationship exists between triglyceride level and HDL-cholesterol level hence exact impact of raised levels of triglycerides compared to that of LDL: HDL ratios are as yet unknown. In a healthy person the level of triglycerides should be around 150 mg/dL

High density lipoprotein (HDL) cholesterol is a kind of good cholesterol and removes fatty deposits from inside the blood vessels and takes them to the liver for removal. Low levels of HDL cholesterol therefore increase the risk for heart disease.

12

Other factors leading to increase incidence of heart diseases include following:

High blood pressure. In this case the heart has to work hard to pump blood. Thus high blood pressure strains the heart, damages blood vessels, and increase the risk of heart attack and stroke.

Smoking and diabetes. It is estimated that smoking can doubles the risk of getting heart disease. Stopping smoking is especially important for people with diabetes because both smoking and diabetes narrow blood vessels.

Strategies to Prevent Heart Diseases

Persons at high risk of heart disease can take a number of measures to reduce the cardiovascular risk. These are:

- Maintaining strict diet control
- Cutting down on saturated fats
- Consuming cholesterol to less than 300 milligrams a day
- Keeping amount of trans fat in to a minimum
- Regular physical activity
- Maintaining a healthy body weight
- Quit smoking

Lowering total LDL cholesterol and lowering blood pressure benefits both, diabetic and non-diabetic patients. Angiotensin converting enzyme inhibitors, β-blocking agents, aspirin and thrombolytic therapy are also effective for the treatment of cardiovascular disease in diabetic patients.

The thiazolidinediones (TZDs) treatment of hyperglycemia in diabetic patients may also reduce the progression of atherosclerosis. These drugs reduce hyperglycemia and improve insulin sensitivity. Treatment of diabetes with TZDs may not be advisable in those diabetic patients showing evidence of congestive heart disease or who are at risk of developing this condition. In another study of randomized double blind clinical trials, a 2.3% incidence of serious heart failure was observed in those patients treated with pioglitazone compared with 1.8% in the control group, although the incidence of myocardial infarct was lower in the pioglitazone group. In yet another trial with pioglitazone or rosiglitazone it was also observed that these medications have an increased incidence of congestive heart disease but there was no overall increased risk of death from cardiovascular disease in the TZD treated groups compared with the controls.

Cerebrovascular Disease

Cerebrovascular disease is a group of brain dysfunctions affected due to poor circulation of blood to the brain, causing limited or no blood flow in affected areas. One of the main reasons of cerebrovascular disease is atherosclerosis. The high level of cholesterol together with inflammation in areas of the arteries in the brain leads the cholesterol to build up in the vessel in the form of a thick, waxy plaque. These plaques can either completely stop the blood flow to the brain or limit its supply depending on the size of the plaques. Furthermore high blood pressure damaging to the blood vessel lining and endothelium, and exposing the underlying collagen, here the platelets aggregate to initiate the repair process which is rarely complete and perfect. Hence if hypertension continues for a long time it can permanently narrow down the blood vessels. Furthermore, their stiffness and deformation make them more vulnerable to fluctuations in blood pressure.

In asymptomatic patients, routine screening for coronary artery disease is not recommended, as it does not improve outcomes as long as coronary vascular disease risk factors are treated.

12

Peripheral Arterial Disease

Peripheral arterial disease (PAD) among diabetic patients occurs due to atherosclerosis. The likelihood of its occurring, especially in diabetic patients of over 50 years, is fairly high (about 30%). Overweight smokers of over 50 with high blood pressure, abnormal blood cholesterol levels, physically inactive and have family history of heart disease are in greater risk of developing this disease. Symptoms may not be very apparent and can easily be overlooked, but certain problems linked with the legs can lead to its possibility. These include:

- Feeling numb or coldness and tingling in the lower part of the legs
- Leg pain during walking and which vanishes after small rest
- Infections or soar on the feet or legs
- Any wound or soar demanding longer healing process
- Peripheral arterial disease can be assessed by Ankle Brachial Index, comparing blood pressure in the ankle with blood pressure in the arm. Magnetic resonance imaging, ultrasound or angiogram can confirm the disease.

The treatment is about the same as needed to control diabetes, heart diseases and stroke. In some cases, surgical procedures such as angioplasty or arterial bypass graft are used to treat this disease.

Foot Diseases

Diabetic foot disease is rather uncommon but equally important as lack of care for this complication can lead to leg amputation. In this disease mostly those patients suffer who either ignore caring their feet or take measure at such a stage when the problems enter in an incurable phase to the extent that amputation remains the only measure to save life. In many cases foot disease is preventable if the precautions mentioned below are taken.

As said earlier, diabetic patients, due to poor defence, are more prone to microbial infection. Additionally they may already have developed neuropathy or peripheral vascular disease which can be contributory to enhancing problems associated with the foot disease such as foot ulceration due to cuts and abrasions.

Foot infection can be classified into localized and generalized infection. In localized foot infection redness and swelling are confined to a specific area of the foot and can be more easily treated by suitable antibiotics. If the ulceration is associated with neuropathy, the patient may not feel the pain or discomfort. The generalized type foot infection may require longer treatment including surgical interventions due to deep tissue infection while the localized type may be cured rapidly.

Charcot's Arthropathy is a complication of diabetic neuropathy in which usually one foot becomes affected. In this complication the foot bone becomes weaker and fragile and can fracture relatively easily even without trauma. As it is associated with diabetic neuropathy the patient usually does not realized any pain and continue to use feet to walk or even to run. This can lead to much damage to the foot bone and resulting foot deformity and if ulceration and infection continue, it can almost certainly lead to foot amputation. It is therefore very important to make the diagnosis early at the acute phase of Charcot's arthropathy because appropriate treatment can minimize deformities and reduce associated morbidity.

It is important to note that Charcot's arthropathy usually develops slowly and without notice and suddenly it flares up inducing swelling and warm to touch foot.

Risk Factors for Diabetic Foot
- Poor glycemic control
- Degree of neuropathy
- Peripheral vascular disease
- Inability of the person to take good self care
- Past history of foot ulceration

Foot Care Instructions
- Always walk with suitable footwear.
  Avoid wearing very tightly fitted shoes.
- Keep feet hygienically clean. While drying it is important to properly dry the space between the two fingers. Most fungi try to grow in this region due to moisture.
- Take good care of nails and keep them trimmed so that microbes cannot grow between the overgrown nails and skin.

- Check feet everyday and look for any cuts or wound, sign of infection, swelling, changes in shape and color.
- Consult doctor immediately for any foot infection.
- Diabetics should not walk barefoot and refrain from walking on very hot or very cold land.
- Do not use any sharp instrument on skin for any reason such as removal of corn etc.

12

CHAPTER 13

Miscellaneous Complications

Slow Wound Healing

It is well known that diabetic patients' wound does not heal up as rapidly as the same kind of wound in non-diabetic persons. Several reasons for it are briefed below.

Poor blood circulation. It happens due to atherosclerosis. Poor circulation reduces the amount of oxygen and healing materials reaching at the site of wound and hence delay occurs in the healing process.

Nerve damage. It occurs because of neuropathy. Its effect on feet can be loss of sensation and numbness and for this reason the patient may not recognize any cuts and wounds as soon as they occur and the problem can continue for some time without treatment which subsequently can become infected. Also callus can be produced due to unfit shoes. The callus usually puts pressure on the deeper layers of the skin which can be converted into blood blister. The centre of callus, being softened by blister, then can convert to an open sore.

Impairment of immune system. Body's Immune systems play important roles in the healing up of the wounds. As diabetic patients have their immune system being impaired the healing process automatically becomes slow. Furthermore the wound can get infected by pathogenic organisms. As the immune systems play major roles in combating the microbial infection, its impairment delays the destruction of infective agents and hence delayed wound healing.

Periodontal Disease

Periodontal disease is the disease of gum, the tooth-supporting tissues and the teeth themselves. Amongst various forms of periodontal diseases there are two types which have most been studied: one is the mild form of gum disease in which the gum is inflamed known as gingivitis. This may cause the gum to bleed during brushing or flossing. The other kind is more serious known as periodontitis. When gingivitis remains untreated for some time, it can lead to the mild form of periodontitis. At this stage erosion of jaw bone occurs around the teeth. The next stage of periodontitis

Diabetes: A Comprehensive Treatise for Patients and Care Givers,
by Shamim I. Ahmad and Khalid Imam. ©2014 Landes Bioscience.

is of the severe form which is characterized by significant tissue and bone loss around the teeth.

Symptoms of Periodontal Disease

Symptoms of periodontal disease include:

- Persistent bad breath
- Swollen, red or tender gums
- Gum bleeding specially during brushing and flossing
- Presence of pus between teeth/gums
- Loosening of teeth
- Teeth alignment impaired

The first clinical manifestation of periodontal disease is the appearance of periodontal pockets, where microbial colonization occurs. The etiology of periodontal disease is multifactorial caused by the interactions between a plethora of micro-organisms, a host with some degree of susceptibility and the environmental factors. Current studies suggest that diabetes is associated with an increased incidence and progression of periodontitis and periodontal infection is associated with poor immune system due to improper glycemic control.

Exact reasons for increased incidence of periodontal disease amongst diabetic patients have not been fully worked out but possible reasons have been put forward which include:

- **Loss of immune system**: as has been discussed above that the impairment of immune system can delay the wound healing process same may be applicable with periodontal disease that the gum infection may not be healing swiftly in diabetes—hence the increased risk of disease.
- **Blockage or damage to capillaries supplying blood to gum**.
- **Excess lipid in blood of diabetic patients** which may be producing chemicals which inflame the gum.

Preventive Measures for Periodontal Disease

For precise analysis of the infective agents appearance of the depth of periodontal pockets is important to be found by clinical examination with periodontal probe plus with X-ray imaging and microbiological techniques.

Occurrence of periodontal disease can be prevented by taking good care of mouth specially teeth. Teeth should be brushed with full care and attention with good tooth brush and may be twice daily, once before going to bed. As periodontal disease is one of the least unwanted complications in diabetes, ironically this disease has not been drawing as much attention as other diabetic complications. .

Psychological Impact of Diabetes

Diabetes in different countries and in different societies is taken differently, ranging from total carelessness to serious treatments including care for psychological effects. In those societies where diabetes is not taken seriously, there are multiple reasons for it including the illiteracy, ignorance, poverty, shortage of specialized doctors and many more. But in those countries where the health care is of superior quality, even there the care and degree of awareness of psychological effects of diabetes may be very limited, mainly due to limited health funding and/or shortage of specialized clinicians and nurses for this complication.

No doubt that diabetes can have significant impact on the psychology of patients. For them diabetes not only can change their lifestyle but also can bring about a large number of feelings and emotion that should be considered while the treatment is going on. Some of the psychological impacts of diabetes are discussed below.

Depression

Depression in diabetes is common and can lead to devastating consequences. There may be several reasons including difficulties in entering in new phase of life. Erectile dysfunction can be another major cause of depression and must be handled with great care by the specialists. Therefore, clinicians and the families should keep a close watch for any sign of depression. The earlier the psychological support provided, the better is the chances to avert any depression. In fact the patients may be made aware that "the diabetes can be a blessing in disguise" for them. Also the patient should be directed towards positive thinking about the disease and should be made aware that it should not be difficult to manage to live a normal and perhaps healthier life with diabetes.

If the family and doctor will fail to handle the serious kind of depression the patient has entered in, (s)he should immediately be referred to professional counseling and assistance to help deal with the effects of the disease. Prolonged depression can lead to weakened immunity and lack of physical activity—further complicating the diabetic control and the treatment.

Anger

Anger is a normal phenomenon in human nature but diabetic patients can become more angry due to the reasons that they have to suffer from a number of prohibitions, e.g., not to be able to eat what they like and will have to continue various diabetes controls measures throughout their life and so on. Also fear may exist for heart attack, kidney disease or blindness or even limb amputation. Regular anger of this kind may lead to anxiety or even depression, which should be helped to be controlled by specialists. Again a self control, positive thinking and support from the family and care givers may be employed to suppress this issue.

Hiding Disease

Hiding disease can be a culture problem as in some countries, such as in South Asia it is considered unwise if the neighbor and friends learn about the disease. In some cases (especially for a female) it can become difficult to find a marrying partner if the person is suffering from diabetes. The disease will be considered as a life long health burden on the groom's family. In such cases the patients are doubly hurt and therefore psychological impacts can be serious.

We hope that the society globally will change their attitude by realizing that no person falls ill from his/her own choice and consider it as the part of life and any body at one time or another may fall ill. In fact we should be more sympathetic with the patients suffering from chronic diseases such as diabetes and allow them to talk openly and frankly about their disease without being embarrassed.

Embarrassment

In the daily life of diabetic patients there comes time when they have to carryout certain function such as blood test or insulin injection in the company of friends or even in a public place such as restaurant or traveling in a train or a bus. Some patients can take it as a normal procedure but others can feel embarrassed. Such embarrassment is a common psychological effect of the disease. Diabetics should not feel embarrassed in such situations, instead they should feel free without any embarrassment to declare that they are suffering from diabetes, taking insulin and are not allowed to eat certain food on medical ground.

Acceptance

After the initial shock, the diabetic patients will ultimately come to accepting the disease. In this phase of acceptance the patients is satisfied to think that there is no body in this world who never suffer one or the other kind of disease. Nature has given him to share diabetes with other millions in the world. As soon as the acceptance comes all the above phenomena (embarrassment, anger and depression) will minimize. Also in this phase the patient will come to terms with their disease and will realize that they have to change to a healthy lifestyle to lead a full, happier and productive life. Knowing that they can change the way they can take care of themselves much better which will help them to live longer and delay diabetic complications.

13

Finally, diabetes related psychological and social problems can impair the ability of families and friends to carry out the patient's care and may lead to compromised health status. Therefore, it is important for the clinicians to assess psychosocial status of the patients in efficient manners so that the referral for appropriate services can be accomplished timely.

13

CHAPTER 14

Management of Diabetes

A golden rule for managing diabetes is: Keep a close watch on the blood sugar levels and try to keep them as much in the normal range as possible by diet, exercise and antidiabetic medication.

The major objectives of proper diabetes management are to alleviate the symptoms, achieve good glycemic control and prevent the micro and macrovascular damage. Data from the UKPDS, DCCT and ADVANCE studies demonstrate the substantial impact of good glycemic control on vascular complications. It is estimated that a 1% decrease in $HbA1_C$ result in 21% reduction in diabetes related deaths, 37% reduction in microvascular complications and 14% reduction in myocardial Infarction.

Optimum Level of Good Glycemic Control

To avert the diabetes complication, it is prudent to achieve a good glycemic, blood pressure and cholesterol control. See Tables 14.1 and 14.2 for the recommended targets as set forth by American Diabetes Federation (ADA) and International Diabetes Federation (IDF).

Management of Type 2 diabetes demands a thorough approach which includes achievement of good glycemic control, diabetes education, minimization of cardiovascular risk, an emphasis on life style modification, and avoidance of drugs which can aggravate glucose or lipid metabolism, and screening for diabetes complications. Comprehensive diabetes management can delay the progression of diabetic associated complication and maximize the superlative quality of life.

14

Table 14.1. Glycemic Targets: Current Recommendations

	*ADA
Pre prandial glucose mg/dL	90–130
Post prandial glucose mg/dL	<180
$HbA1_C$ %	<7.0

* American Diabetes Association

Diabetes: A Comprehensive Treatise for Patients and Care Givers,
by Shamim I. Ahmad and Khalid Imam. ©2014 Landes Bioscience.

Table 14.2. ADA Recommended targets for Cholesterol and B.P.

LDL-C	<100 mg/dL <70 mg/dL with many CV risk factors
Triglycerides	<150 mg/dL
Total Cholesterol	<200 mg/dL
HDL-C	Man: >40 mg/dL Woman: >50 mg/dL
Blood pressure	<130/80 mm Hg

Diabetes Education

Diabetes education is one of the most important obligations on the part of a clinician. Patients and their families are the best persons to manage the disease which is affected so markedly by daily fluctuations in environmental stress, exercise, diet and infections. Education should include explanations by the physician or diabetes educators of the potential acute and chronic complications of diabetes and how they can be recognized early and prevented or treated. Self monitoring of blood glucose should be emphasized, especially for Type 1 diabetes and insulin requiring Type 2 diabetes, and instructions must be given on proper testing and recording of the data.

14

Patients taking insulin should be taught how to adjust the insulin dose for the carbohydrate content of a meal. Strenuous exercise can precipitate hypoglycemia and patients must therefore be taught to reduce their insulin dosage in anticipation of the outcome of strenuous activity or to take supplemental carbohydrates.

Dietary Control and Medical Nutrition Therapy

Dietary restriction is one of the most difficult tasks to be carried out but is the cornerstone of diabetes management and a well-balanced diet remains a cornerstone of therapy. Dietary control can be achieved through medical nutrition therapy; a process by which the nutrition prescription is tailored for people with diabetes is based upon medical, lifestyle and personal factors. The American Diabetes Association recommends about 45–65% of total daily calories should be taken in the form of carbohydrates, 25–35% in the form of fat (of which <7% are from saturated fat) and 15–35% in the form of protein. Type 1 and Type 2 diabetic patients, taking insulin, must be trained for "carbohydrate counting," so that they can administer their insulin bolus after each meal based on its carbohydrate content. In obese diabetics, an additional target is weight reduction by calorie count restriction.

The current recommendations are to limit cholesterol intake to 300 mg daily, and individuals having LDL cholesterol more than 100 mg/dL should limit dietary cholesterol to 200 mg/day. Diabetic patient should also be instructed to take daily fibers in their diet. Fibers have good beneficial effects on cholesterol in the body.

High protein intake is known to enhance kidney disease in patients with diabetic nephropathy; for them a reduction in protein intake to 0.8 kg/day is recommended.

Exercise

There is a common understanding that regular exercise provides immense benefit to our health; more so to diabetic patients. Exercise improves the insulin sensitivity and diabetes control. It also helps to reduce the diabetic complications specially the heart attack or stroke. Also the control of good blood pressure can be achieved by exercise. It brings improvement in blood cholesterol and fat and hence reduces the chances of developing insulin resistance. Also exercise is a must for obese (also for non-obese) diabetic patients. With this huge number of benefits exercise is being increasingly promoted as part of the therapeutic regimen for diabetes.

Based on the literature, if completely sedentary and under active individuals participate in moderate physical activity of 30 minutes a day, they would obtain at least a 30% risk reduction related to diabetes complications. The American Heart Association and the American Diabetes Association recommend at least 150 minutes of moderate intense aerobic activity or at least 90 minutes of vigorous aerobic exercise per week should be carried out. The activity should be distributed over at least five days each week, with no more than 2 consecutive days of inactivity. Before embarking to exercise regime it is important to consult your doctor or the diabetic nurse for the type and extent of exercise needed. Brisk walking, cycling, swimming and running are the kind of good exercises.

Further tips for the exercise are given below:
- Bring variations in exercise regime.
- Socialize or go for the exercise with a friend. This will maintain routine exercise regime and increase motivation.
- Set a goal, especially those important for reducing and maintaining weight in obese diabetics.
- Use the "warming up" criterion which means that start with slow exercise for about 5 minutes and then increase the duration gradually to tune up the body. This will allow heart beat to adapt to enter in with faster rate without causing any harm. Likewise use the "cool down" criterion (again for about 3–5 minutes) so as to bring heart to a normal state of beat. The faster it will come down to normal state the better one's heart is.

Note added from the experience of the author of this book; on an exercise bike with a moderate cycling speed of 15 Km per hour, with median pedaling pressure, and for about 45 minutes with a distance covered about 10 Km and calories loss of about 250 can affect the blood glucose level (measured by a glucometer) to be reduced from 10 mmol/L to as low as 5 mmol/L.

Obesity and Weight Loss

More than 80% of cases of Type 2 diabetes can be attributed to obesity, which may also account for many diabetes-related deaths. In those people the treatment regime should include efforts to reduce weight. The treatment to reduce hyperglycemia by insulin, thiazolidinediones and sulphonylureas may increase weight; however, metformin, dipeptidyl peptidase IV inhibitors and glucagon-like peptide-1 receptor agonists may in fact reduce weight.

Despite the clear benefit of weight loss, ironically only a small percentage of Type 2 diabetic patients remain able to attain and maintain substantial weight loss. This difficulty results from limited success in long-term adherence to calorie-restricted diets. However, impressive success has been reported with intensive dietary intervention.

Pharmacotherapy for weight loss may be effective in Type 2 diabetic patients and generally is associated with high dropout rates due to its side effects and is not recommended as primary therapy for diabetes. Orlistat, a pancreatic lipase inhibitor, is the only FDA approved drug for the long term management of obesity. Other weight-losing drugs such as rimonabant and sibutramine have been withdrawn from the market because of increase suicidal thoughts and cardiovascular side effects. Exercise and diet control (described above) still remains the best measures for weight control.

In July 2012, FDA approved combination of phentermine and topiramate (Qsymia) for the long term management of obesity in conjunction with diet and exercise.

Bariatric Surgery for Obesity

Bariatric surgery may be considered for adults with BMI >35 kg/m^2 when diabetes or associated co-morbidities are difficult to control with lifestyle and pharmacologic therapy. Gastric bariatric surgery results in the largest degree of sustained weight loss and improvements in blood glucose control and should be considered in morbidly obese diabetics. Bariatric surgery has been shown to lead to near or complete normoglycemia in more than 55% to 95% of patients with Type 2 diabetes, depending on the surgical procedure.

14

Bariatric surgery is achieved via reducing the size of the stomach with an implanted medical device or through removal of a portion of the stomach or by resecting and re-routing the small intestines to a small stomach pouch.

Cardiovascular Risk Factors Management

In addition to glycemic control, it is important to minimize the risk for cardiovascular diseases in diabetic patients. These include complete smoking cessation, although controversial for its benefit, use of aspirin (as secondary prevention), blood pressure control and reduction in serum lipids. These should be given the top priorities amongst all types of diabetic patients.

Vaccination

Influenza and pneumonia are common, preventable infectious diseases associated with high mortality and morbidity in the elderly and in people with chronic diseases. Diabetic patients should receive influenza vaccination yearly and pneumococcal vaccination repeating it once after 65 years of age if the initial vaccination was prior to age 65.

Self Monitoring of Blood Glucose

Whether suffering from Type 1 or Type 2 diabetes, in writer's opinion, it is equally important to keep on self monitoring of the levels of blood glucose on a regular basis and maintaining a good record of blood glucose level together with the food consumed and the dose of medications taken, specially, insulin.

14

CHAPTER 15

Pharmacological Treatment of Diabetes Mellitus

Diabetes care is best provided by a multidisciplinary team of health professionals with expertise in diabetes, working in collaboration with the patient and family. The goals in caring for patients with diabetes mellitus are to eliminate symptoms and to prevent, or at least slow down, the development of complications. Microvascular (i.e., eye and kidney disease) risk reduction is accomplished through control of glycemia and blood pressure; macrovascular (i.e., coronary, cerebrovascular, peripheral vascular) risk reduction, through control of lipids and hypertension, smoking cessation, and aspirin therapy, and metabolic and neurologic risk reduction, through control of glycemia.

Management should include the following:
- Appropriate goal setting
- Dietary and exercise modifications
- Medications
- Appropriate self-monitoring of blood glucose (SMBG)
- Regular monitoring for complications
- Laboratory assessment

Pharmacological Interventions

Pharmacological interventions include use of oral medications and injectable therapy.

A. Oral Medications

After a successful initial response to oral therapy, Type 2 diabetic patients usually fail to maintain the $HbA1_C$ target level (<7%). They usually are increased at a rate of 5 to 10% of the 7% per year. Also studies have shown that 50% of patients originally controlled with a single drug required the a second additional drug after three years and by 9 years 75% of patients needed multiple therapies to achieve the target of $HbA1_C$.

15

Diabetes: A Comprehensive Treatise for Patients and Care Givers, by Shamim I. Ahmad and Khalid Imam. ©2014 Landes Bioscience.

Early initiation of pharmacologic therapy is associated with improved glycemic control and reduced long-term complications in Type 2 diabetes. Drug classes used for the treatment of Type 2 diabetes include the following:

Sulfonylureas

Sulfonylureas (eg, glyburide, glipizide, glimepiride) are insulin secretagogues that stimulate insulin release from pancreatic β cells and probably have the greatest efficacy for glycemic lowering of any of the oral agents. However, that effect is only short-term and quickly dissipates. Sulfonylureas may also enhance peripheral sensitivity to insulin secondary to an increase in insulin receptors or to changes in the events following insulin-receptor binding.

Sulfonylureas are indicated for use as adjuncts to diet and exercise in adult patients with Type 2 diabetes mellitus. They are generally well-tolerated, with hypoglycemia the most common side effect.

The first-generation sulfonylureas are acetohexamide, chlorpropamide, tolazamide, and tolbutamide; the second-generation agents are glipizide, glyburide, and glimepiride. The structural characteristics of the second-generation sulfonylureas allow them to be given at lower doses and as once-daily regimens.

Table 15.1 gives a comparison between different sulphonylureas and other oral medicatios.

Hypoglycemia is the most common side effect and is more common with long-acting sulphonylureas such as chlorpropamide, glyburide/glibenclamide. Elderly patients are more prone to develop hypoglycemia, therefore short acting sulphonylureas are preferable, such as gliclazide and glipizide.

Drug-induced hypoglycemia is most likely to occur in the following circumstances in older patients and may be a limiting factor for use of these drugs in older adults:

- Exercise
- Missed meals
- Impaired hepatic or renal functions
- Alcohol abuse
- Drug therapy with salicylates, gemfibrozil, warfarin etc.
- Poor left ventricular function
- Porphyria

Weight gain may occur in some patients using sulfonylurea. Hypersensitivity reaction to sulphonylureas may occur in the first 6–8 weeks of therapy mainly of allergic skin reaction which can progress albeit rarely to erythema multiform, exfoliative dermatitis, fever and jaundice.

Table 15.1. Oral Medications for T2DM

Drug	Strength	Daily Dosage
Sulfonylureas		
Glimepiride	1, 2, 3, 4 mg	1–8 mg once a day
Glibenclamide/ Glyburide	5 mg	2.5–20 mg single or two divided dosages
Gliclazide	80 mg	80–320 mg two divided doses
Gliclazide MR*	30, 60 mg	30–120 mg once a day
Glipizide	5, 10 mg	2.5–20 mg two divided doses
Meglitinide		
Repaglinide	0.5, 1, 2 mg	0.5–4 mg thrice a day
Biguanides		
Metformin	500, 850, 1000 mg	1000–2500 mg two to three divided doses
Thiazolidinediones		
Pioglitazone	15, 30, 45 mg	15–45 mg once a day
Rosiglitazone	4, 8 mg	4–8 mg once a day
α-Glucosidase Inhibitors		
Acarbose	50–100 mg	25–100 mg three times a day
Dipeptidyl Peptidase 4 (DPP-4) Inhibitors		
Sitalgliptin	100 mg	**100 mg once a day
Vildagliptin	50 mg	50–100 mg /day
Saxagliptin	5 mg	5 mg once daily
Sodium Glucose Co-Transporter Inhibitor 2		
Canagliflozin	100 mg	100–300 mg once a day

* Modified Release, ** 50 mg/day if creatinine clearance 30–50 ml/min and 25 mg/day if clearance <30 ml/min

15

Glinides

Meglitinides (eg, repaglinide, nateglinide) are much shorter-acting insulin secretagogues than the sulfonylureas are, with preprandial dosing potentially achieving more physiologic insulin release and less risk for hypoglycemia.

Meglitinides can be used as monotherapy; however, if adequate glycemic control is not achieved, then metformin or a thiazolidinedione may be added. Meglitinides may be used in patients who have allergy to sulfonylurea medications. They have a similar risk for inducing weight gain as sulfonylureas do but possibly carry less risk for hypoglycemia. One another advantage of repaglinide is that it can be given in renal failure patients because primary route of excretion of its metabolite is through hepatobiliary tree.

Biguanides

Metformin is the only biguanide in clinical use. Another biguanide, phenformin, was taken off the market in the United States in the 1970s because of its risk of causing lactic acidosis and associated mortality (rate of approximately 50%). Metformin has proved effective and safe. As with other oral antidiabetic drugs, lactic acidosis during metformin use is very rare and is associated with concurrent comorbidity.

Metformin is a insulin sensitizer and lowers basal and postprandial plasma glucose levels. Its mechanisms of action differ from those of other classes of oral antidiabetic agents; metformin works by decreasing hepatic gluconeogenesis production. It also decreases intestinal absorption of glucose and improves insulin sensitivity by increasing peripheral glucose uptake and utilization. Unlike oral sulfonylureas, metformin rarely causes hypoglycemia.

Patients on metformin have shown significant improvements in hemoglobin A1c and their lipid profile, especially when baseline values are abnormally elevated. In addition, metformin reliably facilitates modest weight loss. In the UKPDS, it was found to be successful at reducing macrovascular disease endpoints in obese patients.

Absorption of vitamin B_{12} appears to be reduced during long-term metformin therapy, but the serum vitamin B_{12} levels usually remain in the normal range. However, periodic screening with serum vitamin B_{12} levels should be considered, especially in patients with symptoms of peripheral neuropathy.

Renal dysfunction (serum creatinine >1.5 mg/dL in man and >1.4 mg/dL in women or creatinine clearance <30 ml/min) is considered a contraindication to metformin use because it may increase the risk of lactic acidosis, an extremely rare (less than 1 case per 100,000 treated patients) but potentially fatal complication. Its use should be avoided in acute medical conditions like ketoacidosis, sepsis, hepatitis, respiratory failure, heart failure. It should be withdrawn one day prior to giving radio contrast agent or general anesthesia and restart when renal function returns to normal or at least three days after giving these agents.

Thiazolidinediones (TZDs)

TZDs (e.g., pioglitazone, rosiglitazone act as insulin sensitizers. They must be taken for 12–16 weeks to achieve maximal effect.

These agents are used as monotherapy or in combination with sulfonylurea, metformin, meglitinide, DPP-4 inhibitors, GLP-1 receptor agonists, or insulin. These antidiabetic agents have shown to slow down the progression of diabetes, particularly in early disease.

When used as monotherapy, these drugs lower $HbA1_C$ by about 0.5–1.4%. When used in combination with insulin, they can result in a 30–50% reduction in insulin dosage, and some patients can come off insulin completely.

The mechanism of action of thiazolidinediones is not fully understood but likely that it increases insulin sensitivity by acting on adipose tissues, muscle and liver to increase glucose utilization and decrease glucose production.

Controversy exists for an increased risk of myocardial infarction with rosiglitazone but it is not the same as with pioglitazone. Certain adverse side effects of thiazolidinediones include weight gain, edema, anemia and fracture risk at atypical site (forearm). Recently increased risk of bladder cancer has also been reported after one year of pioglitazone therapy. This group of drug should be avoided in patients with advanced forms of congestive heart failure, liver failure and in pregnancy and lactation because of lack of safety data.

15

Alpha Glucosidase Inhibitors

Acarbose and miglitol are included in this group. These agents delay sugar absorption and help to prevent postprandial glucose surges. Alpha-glucosidase inhibitors prolong the absorption of carbohydrates by blocking the enzyme α-glucosidase in the small intestine, but their induction of flatulence greatly limits their use. They should be titrated slowly to reduce gastrointestinal (GI) intolerance. A fundamental difference between acarbose and miglitol is in their mechanism of absorption. Very little acarbose (about 2%) crosses the microvillar membrane whereas miglitol has a structural similarity with glucose and is more absorbable.

Acarbose should be avoided in inflammatory bowel disease, renal and hepatic impairment, pregnancy and lactation.

Dipeptidyl Peptidase-4 Inhibitors

Several dipeptidyl peptidase 4 inhibitors are available in the market such as sitagliptin, vildagliptin, saxagliptine and linagliptin. DPP-4 inhibitors are a class of drugs that prolong the action of incretin hormones. DPP-4 degrades numerous biologically active peptides, including the endogenous incretins GLP-1 and glucose-dependent insulinotropic polypeptide (GIP). DPP-4 inhibitors can be used as a monotherapy or in combination with metformin or a TZD. They are given once daily and are weight neutral.

Sodium-Glucose Co-Transporter Inhibitors (SGLT2)

Sodium-dependent glucose co-transporters (or sodium-glucose linked transporter, SGLT2) are a family of glucose transporter found in the intestinal mucosa of the small intestine. It is exclusively expressed in the proximal renal tubules, accounts for about 90% of the reabsorption of glucose from tubular fluid.

Canagliflozin (Invokana)

In March 29, 2013, canagliflozin became the first SGLT2 inhibitor approved by FDA for the treatment of Type 2 diabetes. Canagliflozin is an inhibitor of sodium-glucose co-transporter 2 (SGLT2), which is responsible for at least 90% of the glucose reabsorption from proximal convoluted tubules of the kidney. Blocking this transporter causes blood glucose to be eliminated through the urine.

Dose of canagliflozin is 100–300 mg/day before meal. It can be used in combination with other oral anti diabetics and insulin. Its beneficial effects includes decreasing weight by a 1.9–3% as well as decreases hemoglobin AIC by 0.57–0.70%, reduces both systolic and diastolic blood pressures, raises HDL cholesterol and causes less hypoglycemia.

Their side effect includes urinary tract infections, genital mycotic infections and is associated with increased urination and episodes of hypotension and hyperkalemia. It may increases LDL cholesterol and cardiovascular side effects have also been reported with in this group of drugs.

Canagliflozin should not be used in the following conditions:
- Pregnancy
- Lactation
- Type 1 diabetes
- Moderate to severe renal diseases

B. Injectable Therapy

Insulin

Insulin is the only option for treating Type 1 and gestational diabetes. Ultimately, many patients with Type 2 diabetes mellitus become markedly insulinopenic. The only therapy that corrects this defect is insulin. Because most patients are insulin resistant, small changes in insulin

15

dosage may make no difference in glycemia in some patients. Furthermore, because insulin resistance is variable from patient to patient, therapy must be individualized in each patient. Insulin is a reasonable choice for initial therapy in patients who present with symptomatic or poorly controlled diabetes and preferred in those patients with HbA1$_C$ >8.5% or with symptoms of hyperglycemia despite therapy with metformin and lifestyle intervention.

Clinical Indications for Insulinization

Following are the indications to initiate insulin therapy:
- Failure to oral medications
- Critically ill conditions
- Diabetes ketoacidosis
- Hyperglycemic hypomolar state
- Liver and kidney failure
- Severe cardio-respiratory dysfunctions
- Pregnancy and lactation

General Principles of Insulin Therapy

Initial therapy should begin with diet, weight reduction, and exercise, which may induce normoglycemia if compliance is optimal. Metformin therapy (in the absence of contraindications) may be initiated, concurrent with lifestyle intervention, at the time of Type 2 diabetes diagnosis.

Clinical features that, if present in a patient with diabetes at any age, suggest the need for insulin therapy include marked and otherwise unexplained recent weight loss (irrespective of the initial weight), a short history with severe symptoms, and the presence of moderate to heavy ketonuria. Diabetic ketoacidosis at first presentation usually indicates that the patient has Type 1 diabetes and will require lifelong insulin treatment. However, some patients with Type 2 diabetes, especially in the Afro-Caribbean populations (so-called "Flatbush diabetes") may present with ketoacidosis.

For many patients with Type 2 diabetes, a basal supplement is often adequate for good glycemic control as endogenous insulin secretion will control the post-prandial excursions. Some patients with Type 2 diabetes may require additional pre-meal boluses, similar to Type 1 diabetics.

The use of premixed insulin is not generally recommended for patients with Type 1 and gestational diabetes since these offer little glycemic advantage compared with adequately titrated basal and bolus insulin. Many patients with Type 2 diabetes can use pre-mixed preparations with reasonable effect.

15

Table 15.2. Pharmacokinetics of Conventional Insulin

Types	Onset	Peak	Duration
Regular	30–60 min	2 h	4–6 h
NPH	2–4 h	4–6 h	12–18 h

Conventional Insulin

Conventional therapy usually involves one to three daily injections. The types of insulin and the number of injections and doses are determined by quantity and quality of foods, physical activity and other factors. Regular and Neutral Protamine Hagedron (NPH) are conventional insulin. Premixed formulation 30/70 (30% regular and 70% NPH) is also available in the market.

Regular Insulin

Regular insulin is short-acting soluble crystalline zinc insulin whose effect appears within 30 min after subcutaneous injection and lasts 4–6 hours. See Table 15.2.

NPH (Neutral Protamine Hagedorn) Insulin

NPH insulin is intermediate acting insulin whose onset of action is delayed by combining two parts soluble crystalline zinc insulin with one part protamine zinc insulin. Its onset of action is delayed to 2–4 hours, and its peak response is generally reached in about 4–6 hours. Because its duration of action is often <24 hours (with a range of 10–20 hours), most patients require at least two injections daily to maintain a sustained glycemic effect.

Insulin Analogues

An insulin analog is a type of insulin that has been chemically modified to either act faster or slower than the type of insulin naturally made by the body.

The new insulin analogs, including the rapid-acting analogs (aspart, lispro, glulisine), the long-acting basal analogs (glargine, detemir), and the premixed insulin analog formulations (75% neutral protamine lispro, 25% lispro; 50% neutral protamine lispro, 50% lispro; 70% protamine aspart, 30% aspart) have been formulated to allow for a closer replication of a normal insulin profile.

Rapidly-Acting Insulin Analogues

Insulin lispro, insulin aspart and insulin glulisine are produced by recombinant technology, wherein certain amino acids together with their

15

Table 15.3. Pharmacokinetics of Insulin Analogue

Types	Onset	Peak	Duration
Rapidly Acting Analogs			
Lispro Aspart Glulisine	5–15 min	60 min	2 h
Long Acting Analogs			
Glargine	2–4 h	None	24 h
Detemir	2–4 h	None	20–24 h

positions have been changed. Alteration of the amino acids sequences in these analogs results in formation of monomers when injected subcutaneously in contrast to regular human insulin, whose hexamers require considerably more time to dissociate and absorb. Rapid-acting insulin analogs usually begin working within 15 min after injection, reach peak effectiveness in 30–90 min and have duration of up to 2 to 4 hours (see Table 15.3).

Long-Acting Insulin Analogues

Long-acting insulin analogs have been genetically altered to prolong the release of insulin in the body. There are currently two long-acting insulins that are designed to acts as basal insulin, or the ongoing insulin that is needed to manage normal fluctuations in blood sugar between meals and during sleep.

The first long-acting basal insulin was approved by the FDA in 2000 and goes by the generic name of insulin glargine. The second long-acting insulin was approved in 2005 and is called insulin detemir. Both Lantus and Levemir begin working within about an hour after injection but are released very slowly and evenly into the bloodstream with no significant peak in the action (see Table 15.3). These basal insulins are designed to continue working for up to 24 hours and are usually supplemented with additional injections of rapid-acting or short acting insulin at meals.

Side Effects of Insulin Therapy

Hypoglycemia, Weight Gain and Edema

These are the most frequently occurring side effects of insulin treatment. The common causes of hypoglycemia include missed meals, excessive insulin dosage, erratic meal timing and unplanned exercise, renal impairment, cortisol deficiency, hypothyroidism etc. Sodium and fluid retention is a common occurrence after insulin therapy. Weight gain may be secondary to increased lipogenesis. Insulin's sodium retaining effect on the kidney could be the best explanation for causing fluid retention and edema.

Insulin Allergy

This problem was very common with beef insulin usage and now the problem of insulin allergy has been greatly reduced by the introduction and widespread use of human insulin. There have been case reports of successful use of insulin lispro in those rare patients who have a generalized allergy to human insulin or insulin resistance due to a high titer of insulin antibodies.

Insulin Infusion Pump

The insulin pump is a device for continuous insulin delivery. An insulin pump is composed of a pump reservoir similar to that of an insulin cartridge, a battery-operated pump, and a computer chip that allows the user to control the exact amount of insulin being delivered. This device is about the size of a standard communications beeper and attached to a thin plastic tube that has a soft cannula at the end through which insulin passes. This cannula is inserted under the skin, usually on the abdomen. The cannula is changed every two days. The tubing can be disconnected from the pump while showering or swimming. The pump is used for continuous insulin delivery, 24 hours a day. The amount of insulin is programmed and is administered at a constant rate (basal rate). Often, the amount of insulin needed over the course of 24 hours varies depending on factors like exercise, activity level, and sleep.

The insulin pump allows the user to program many different basal rates to allow for variation in lifestyle. In addition, the user can program the pump to deliver a bolus (large dose of insulin) during meals to cover the excess demands of carbohydrate ingestion.

Regarding continuous subcutaneous insulin infusion pump's clinical efficacy, several studies have shown its clinical superiority over multiple daily insulin injections in terms of HbA1$_C$ control and also concluded that pump, using rapid-acting analogs, is more beneficial in controlling postprandial hyperglycaemia and HbA1$_C$ than the pump using regular human insulin.

More Recent and Future of Diabetes Treatment

Glucagon-Like Peptide-1 Analogues

Glucagon-like peptide-1 (GLP-1) is an endogenous incretin that stimulates insulin secretion, suppresses glucagon secretion, and delays gastric emptying. Current incretin-based analogs approved for Type 2 diabetes include exenatide (Byetta) and liraglutide (Victoza). These agents stimulate pancreas to produce more insulin, reduce the amount of glucose being produced by the liver, reduce the rate by which stomach digest food and act on the brain to cause a feeling of fullness.

Exenatide

It is a synthetic exendin-4 which is a naturally occurring component of the Gila monster saliva. It is resistant to DPP-IV degradation and therefore exhibits a prolonged half-life. Exenatide is dispensed as two fixed-dose pens (5 μg and 10 μg) and administered subcutaneously twice daily just before or within one hour of morning and evening meals. Patients should be prescribed the 5 μg pen for the first month and, if tolerated, the dose can be increased to 10 μg twice a day. The drug is not recommended in patients with glomerular filtration rate <30 mL/min.

Clinical trials, adding exenatide therapy to Type 2 diabetic patients already taking metformin or sulfonylurea or a thiazolidinedione showed that they further lowered the HbA1$_C$ value by 0.4–0.6%. These patients also experienced some weight loss and hence can suit to obese diabetic patients.

Nausea is the major side effect affecting over 40% of the patients. Patients should seek immediate medical attention if they experience unexplained persistent severe abdominal pain. Other side effects include acute renal failure.

Exenatide Extended Release Formulation

Exenatide extended release formulation (Bydureon) is approved by FDA and NICE in January and February 2012 respectively. It is a long acting drug, as compare with exenatide and liraglutide, and can be given once a week for Type 2 diabetes.

Liraglutide

Liraglutide is a long-acting, stable analog of the natural hormone glucagon-like peptide-1. It is intended to be used in combination with basal insulin and an oral antidiabetic agent such as metformin/sulfonylurea for the treatment of adults with Type 2 diabetes mellitus not achieving adequate glycaemic control with diet and other pharmacological interventions. Liraglutide is administered subcutaneously and the dose ranges from 1.2 mg to 1.8 mg daily. It is albumin bound with a half-life of approximately 12 hours, allowing the drug to be injected once a day. Patients may experience sustained weight loss.

Its side effects are nausea and vomiting found in 10–28% users. There is also an increased incidence of diarrhea. In clinical trials, there were seven cases of pancreatitis in the liraglutide treated group with one case in the placebo group.

Pramlintide

This drug is a synthetic analog of islet amyloid polypeptide or amylin. When given subcutaneously, pramlintide delays gastric emptying suppresses glucagon secretion and decreases appetite. It is approved to be used both for Type 1 and Type 2 diabetes in combination with insulin.

Nausea is the major side effect but hypoglycemia can occur specially in Type 1 diabetics. Mild nausea is more likely during the first weeks after starting pramlintide and usually does not last long. It is very important to start pramlintide at a low dose. The short-acting or premixed insulin doses be reduced by 50% when the drug is started.

Long-Acting Insulin Analogue (Insulin Degludec)

Insulin degludec was developed to cover basal insulin needs in patients with diabetes mellitus, either alone or in combination with bolus (mealtime) insulin and/or oral antidiabetic drugs.

Insulin degludec is an ultralong-acting basal insulin analog that forms soluble multi-hexamers on subcutaneous injection. It has a half-life of 25 hours, which is twice as long as the half-life of currently available basal insulin products, with 42-hour duration of effect.

Insulin degludec acts specifically and gives full effect at the human insulin receptor and its mode of action is the same as that of human insulin, thus giving rise to the same metabolic effects such as cellular glucose uptake, glycogen synthesis and lipogenesis.

The European Commission has granted marketing authorization to this long-acting insulin degludec (Tresiba) and a combination agent containing insulin degludec with insulin aspart (Ryzodeg). But theses have not yet been approved by FDA.

Pancreatic Transplantation

The purpose of pancreas transplantation is to ameliorate Type I diabetes and produce complete insulin independence. The first successful pancreas transplantation in conjunction with a simultaneous kidney transplantation was performed by W.D. Kelly, MD, and Richard Lillehei, MD, from the University of Minnesota in 1966.

About 75% of pancreas transplantations are performed with kidney transplantation (both organs from the same donor) in patients with renal failure who are diabetic. This is referred to as simultaneous pancreas-kidney (SPK) transplantation. About 15% of pancreas transplantations are performed after a previously successful kidney transplantation. This is referred to as a pancreas-after-kidney transplantation.

Single transplant of pancreas should only be considered a therapy in patients who exhibit following characteristics:

- A history of frequent hypoglycemia
- Ketoacidosis
- Clinical and emotional problems with insulin therapy

15

CHAPTER 16

An Overview on Gestational Diabetes Mellitus

Introduction

Gestational diabetes mellitus (GDM) is defined as glucose intolerance of various degrees that is first detected during pregnancy. GDM is a common condition affecting ~7% of all pregnancies. The prevalence may range from 1 to 14% of all pregnancies, depending on the population studied and the diagnostic tests employed.

For more than a century, it has been known that diabetes antedating pregnancy can have severe adverse effects on fetal and neonatal outcomes. As early as in the 1940s, it was recognized that women who developed diabetes years after pregnancy had experienced abnormally high fetal and neonatal mortality. By the 1950s the term "gestational diabetes" was applied to what was thought to be a transient condition that affected fetal outcomes adversely, and then abated after delivery. In the 1960s, O'Sullivan found that the degree of glucose intolerance during pregnancy was related to the risk of developing diabetes after pregnancy.

Pathogenesis

Pregnancy is normally attended by progressive insulin resistance that begins near mid-pregnancy and progresses through the third trimester to levels that approximate the insulin resistance seen in individuals with Type 2 diabetes. The insulin resistance appears to result from a combination of increased maternal adiposity and the insulin-desensitizing effects of hormonal products of the placenta. Placental secretion of hormones such as placental lactogen, progesterone, cortisol, prolactin, and growth hormone are major contributors to the insulin-resistant state seen in pregnancy. The fact that insulin resistance rapidly abates following delivery further supports the hypothesis that the major contributors to this state of resistance are placental hormones.

Peripheral insulin sensitivity during the third trimester decreases to 50% of that seen in the first trimester, and basal hepatic glucose output is

16

Diabetes: A Comprehensive Treatise for Patients and Care Givers,
by Shamim I. Ahmad and Khalid Imam. ©2014 Landes Bioscience.

30% higher despite higher insulin levels. This combination of increased mobilization of glucose, along with decreased sensitivity to insulin, places women at risk of developing diabetes during pregnancy.

Gestational diabetic patients also have an impairment of the compensatory increase in insulin secretion, particularly first-phase insulin secretion. This decrease in first phase insulin release may be a marker for deterioration of β-cell function. There is also a subset of women with GDM who have evidence of islet cell autoimmunity.

Risk Factors for GDM

Risk factors of GDM can be divided into high risk and low risk groups.

High Risk for DGM

Presence of any of the following risks is sufficient to level high risk:
- Age more than 25 years
- Prior delivery of macrosomic Infant (weighing >4 kg)
- Increased body mass index
- Family history
- Prior history of prediabetes
- History of miscarriage
- Ethnic Groups such as South Asians, pima Indians

Low Risk for GDM

Presence of all of the following is required for a low risk group:
- Normal body weight before pregnancy
- Age <25 years
- No history of diabetes in family
- Low risk ethnicity
- No prior poor history of miscarriage

Complications

The detection of GDM is important because of its associated maternal and fetal complications. Treatment with medical nutrition therapy, close monitoring of glucose levels, and insulin therapy if glucose levels are above goal can help to reduce these complications.

There are both fetal and maternal complications associated with GDM.

Fetal Complications

Fetal complications include:
- Macrosomia
- Neonatal hypoglycemia
- Perinatal mortality
- Congenital malformation

16

- Hyperbilirubinemia
- Polycythemia
- Hypocalcemia
- Respiratory distress syndrome

Macrosomia, defined as birth weight >4000 g, occurs in ~20–30% of infants whose mothers have GDM. Maternal factors associated with an increased incidence of macrosomia include hyperglycemia, high BMI, older age, and multiparity. This excess in fetal growth can lead to increased fetal morbidity at delivery, such as shoulder dystocia and an increased rate of cesarean deliveries.

Neonatal hypoglycemia can occur within a few hours of delivery. This results from maternal hyperglycemia causing fetal hyperinsulinemia.

Long-Term Complications to the Offspring
These include:
- Increased risk of glucose intolerance
- Type 2 Diabetes
- Obesity

Maternal Complications
Maternal complications include:
- Hypertension
- Preeclampsia
- Increased risk of cesarean delivery
- Increased risk of Type 2 DM

Diagnosis of GDM

16

Gestational diabetes can be diagnosed by two approaches:
- Two-step approach
- Single-step approach

In two-step approach a glucose challenge test (GCT), a screening test, is performed followed by oral glucose tolerance test (OGTT) if GCT screening is positive. In single-step approach a direct oral glucose tolerance test is performed in high risk group.

Glucose Challenge Test (GCT)
Glucose challenge test is the screening test and generally performed in high risk females at 22nd to 24th weeks of gestational amenorrhea since this is the highest period of insulin resistance. It is done with 50 g of glucose at any time of the day irrespective of the time last meal taken. Following glucose ingestion plasma glucose is checked after an hour. Test considered abnormal if glucose level found more than 140 mg/dL. Those who test positive should undergo OGTT (see Table 16.1).

Table 16.1. Diagnostic Criteria of GDM

	ADA 100 g OGTT		ADA 75 g OGTT		WHO 75 g OGTT	
	mg/dL	mmol/L	mg/dL	mmol/L	mg/dL	mmol/L
Fasting	95	5.3	95	5.3	126	7.0
1-hour	180	10.0	180	10.0	— —	— —
2-hour	155	8.6	155	8.6	140	7.8
3-hour	140	7.8	— —	— —	— —	— —

For the ADA criteria, two or more of the values from either the 100- or 75-g OGTT must be met or exceeded to make the diagnosis of GDM. For the WHO criteria, one of the two values from the 75-g OGTT must be met or exceeded to make the diagnosis of GDM.

Oral Glucose Tolerance Test

Gestational diabetes is diagnosed on the basis of OGTT. There are two methods of oral glucose tolerance test in gestational diabetes. One is with 75 g and other with 100 g of glucose. See Table 16.1 for the diagnostic criteria.

Treatment of Gestational Diabetes

Gestational diabetes is managed in the same way as other types of diabetes are being managed but GDM requires more stringent and tight glycemic control to prevent fetal complications. Corner stone of its management includes medical nutrition therapy, exercise and insulin therapy.

Medical Nutritional Therapy (MNT)

MNT performed under supervision of a trained dietitian. It is important that the food provided should carry adequate nutrition for the mother and fetus, sufficient calories for appropriate maternal weight gain, maintain normoglycemia, and avoid ketoacidosis. In normal-weight women with GDM, the recommended daily caloric intake is 30 kcal/kg/day based on their present pregnant weight. In the second and third trimester most normal weight women require an additional 300 kcal/day. A total of 6 small feedings per day should be encouraged with 3 major meals and 3 snacks. Carbohydrates should account for no more than 50% of the diet, with protein and fats equally accounting for the remainder.

For obese women (BMI >30 kg/m^2), a 30–33% calorie restriction has been shown to reduce hyperglycemia and plasma triglycerides with no increase in ketones.

16

Ketonemia in mothers with diabetes during pregnancy has been associated with lower IQ levels and impaired psychomotor development in their children. Monitoring with prebreakfast ketones measurements is recommended for patients who are on a hypo caloric or carbohydrate-restricted diet

Exercise

Exercise is the second component of initial therapy. In addition to promoting general cardiovascular fitness, exercise increases peripheral insulin sensitivity and hepatic glucose uptake. The major concern relating to exercise in pregnancy is its association with uterine contractions and a theoretical risk of subsequent preterm labor or fetal distress. Many exercise regimens have been designed for pregnant women that have not been associated with these complications. Exercises recommended during pregnancy include the arm ergometer, the recumbent bike, and swimming. There is insufficient evidence to recommend a specific type of exercise to manage GDM. The goal of any regimen is to exercise a minimum of 15 min at least 3 times a week. Women with medical or obstetric complications should be carefully evaluated before recommendations are made for participation in physical activities during pregnancy, and they should be closely monitored.

Pharmacological Intervention

Excellent control (2-hour postprandial glucose level < 120 mg/dL and fasting < 95 mg/dL) is usually obtained in the first 2 weeks of adherence to a diet and exercise regimen. If good control is not achieved within the first 2 weeks, or if 2 values per week exceed the target blood sugar levels, control is unlikely to be reached and insulin therapy should be considered.

Insulin Therapy

Insulin therapy is the corner stone of pharmacological intervention of gestational diabetes. The ideal insulin regimen is not known. A starting dose of 0.7 U/kg/day is often recommended in the first trimester titrating up to 0.9 U/kg/day in the third trimester. It should be noted that this dosage is based on ideal body weight and not actual weight. As is the case for starting insulin therapy in any newly diagnosed individual, the total dose is divided into 2/3 in the morning and 1/3 in the evening. The AM dose is split 2/3 Neutral Protamine Hagedorn (NPH) insulin and 1/3 regular insulin. The PM dose is divided 1/2 regular insulin and 1/2 NPH insulin. The NPH formulation is best given as a bedtime dose. Ultra short acting analogs like aspart and lispro can also be used as bolus insulin. Recently, insulin detemir, a long acting basal analog is also recommended to treat gestational diabetes.

Oral Antidiabetic Drugs

There is limited data available on the safety and efficacy of oral anti-diabetics. Therefore oral agents are not generally recommended to treat GDM.

Self Monitoring of Blood Glucose

Self monitoring blood glucose is now considered the standard of care for pregnancy outcome. Recording blood glucose levels in fasting and two hour post meal (four times a day) confers a positive effect on improving glycemic control. Aim of the monitoring is to achieve the glycemic targets as

- Fasting should be <95 mg/dL or 5.2 mmol/L
- 2-hour postprandial should be <120 mg/dL or 6.6 mmol/L

Obstetrical Management

Optimizing outcomes for women with GDM and their fetuses requires not only careful metabolic management, but also appropriately applied fetal surveillance techniques and thoughtful selection of the most advantageous timing and route of delivery. Whenever possible, these clinical decisions should be based on the highest level of evidence available and should weigh the likelihood and seriousness of both maternal and fetal/neonatal morbidity.

Fetal Monitoring

- All women with GDM should monitor fetal movements during the last 8–10 weeks of pregnancy and report immediately any reduction in the perception of fetal movements.
- Non-stress testing should be "considered" after 32 weeks' gestation in women on insulin and "at or near" term in women requiring only dietary management.
- Ultrasound should be used to detect fetal anomalies in women with GDM diagnosed in the first trimester or with fasting glucose levels >120 mg/dL.
- Amniocentesis to determine fetal lung maturity in preparation for delivery is not necessary in well-dated pregnancies after 38 weeks' gestation.

Timing and Route of Delivery

The presence of GDM is not by itself an indication for cesarean delivery. GDM is not an indication for delivery before 38 weeks' gestation in the absence of evidence of fetal compromise. There are not enough data available to draw definitive conclusions on the following issues:

1. The need for intensified fetal surveillance in women with GDM in good control on diet alone.
2. The role of fetal weight estimation in determining the timing and route of delivery.
3. The optimal modality to predict the presence of fetal macrosomia and excessive/disproportionate fetal growth and the occurrence of shoulder dystocia and its resulting birth trauma.

Neonatal Care

Neonatal Hypoglycemia is commonly encountered due to fetal hyperinsulinemia. Hypoglycemia is defined as a heel-stick blood sugar <35 mg/dL in a full-term neonate. The hypoglycemia typically resolves with feeding of either milk or a glucose solution. If the baby is symptomatic or the hypoglycemia profound, an IV bolus of 10% dextrose is recommended @ 0.25 mg/kg followed by 4–6 mg glucose/kg/min with gradual titration. Blood sugar monitoring is continued every hour until it has stabilized.

The potential sequelae of shoulder dystocia are Erb's palsy, a stretch injury to the brachial plexus, and intrapartum fetal hypoxia should always be considered and managed accordingly.

Respiratory distress is the most serious complication for the neonate. Fetuses affected by GDM are at elevated risk of lung immaturity compared with age-matched controls. Oxygen supplementation, ventilatory support, and surfactant replacement are among the treatments available, and care may require consultation with a neonatologist.

16

Post-Partum Follow-Up

The common course of gestational diabetes is complete resolution in the postpartum period, although it can be several days or weeks before the glucose intolerance completely resolves but women with a history of gestational diabetes have an increased risk of developing diabetes after pregnancy compared with the general population. Therefore it is suggested that at 6-week postpartum, they should undergo a 75 g glucose tolerance test. The diagnosis of diabetes is made if fasting plasma glucose is ≥ 126 mg/dL or 7.0 mmol/L or if the 2-hour level is ≥200 mg/dL or 11.1 mol/L. If glucose levels are normal, reassessment of glycemia should be undertaken at a minimum of 3-year intervals.

Conclusion

The objective of this article is to familiarize the readers with issues in the diagnosis and management of GDM, including some of its complications.

Women with gestational diabetes require expert consultation, guidelines and comprehensive diabetes education. Most women with GDM, however, have a relatively uncomplicated course. Further, these women are often highly motivated to control their diabetes in the interest of the health of their children. With frequent follow-up, careful patient education, and specialist consultation if indicated, physicians and their patients with gestational diabetes can expect a healthy pregnancy outcome.

16

CHAPTER 17

Food and Diabetes

In our daily intake one should take food that contains all three basic components such as carbohydrate, proteins, and fats. Vegetables (for roughage and bulk), fruits (for minerals and vitamins) non-fatty milk, yoghurt (for calcium and unsaturated fat) and fish, chicken or other meat (for proteins) and all these must be taken in balanced quantity. Also it is more appropriate to avoid heavy meals 2–3 times per day and take light meals, may be 3–4 times daily. Wheat, corn and rice, oat, barley etc can supply us carbohydrates and milk and milk products are rich in calcium and hence help in teeth and bone growth and their maintenance. Meat, fish, eggs and pulses etc can supply us proteins and fruits and vegetables can supply us various vitamins, fibers and trace elements including iron which is important for synthesis of hemoglobin. Vegetarians who do not take meat or meat products should take milk and other dairy products and the pulses. In pulses comes various nuts, lentils, moong, peas, chana, masoor etc.

Following is a brief description of the food consumed every day.

Sugar

Sugar in various forms especially sucrose and glucose are important for every individual because they are the most important sources of energy. The various forms of sugars include sucrose, glucose, dextrose, fructose, galactose, lactose and maltose and several others. They are present in different natural food such as fructose is rich in fruits and in certain vegetables and lactose in milk.

The irony of modern society is that we are consuming too much sugar for example in soft drink (330 ml containing up to 36 g) chocolates (40 gram of bar having 16 g) than what we need. If we look into the packet of ready-made food in a super market we may find that almost every product there is with the label, you will find the note "carbohydrate (x %) of which sugar (y %). Even some products may label "sugar free" but may have added "grape juice" which is rich in glucose. Some products such as bread, sauces, cereals, fruit juice and the ready meals, where one may not expect to find sugar, may contain sugar in different amount. For a non-diabetic person excessive sugar can lead to obesity (now becoming a pandemic globally),

Diabetes: A Comprehensive Treatise for Patients and Care Givers,
by Shamim I. Ahmad and Khalid Imam. ©2014 Landes Bioscience.

including in young children, and subsequently possible diabetes and heart disease. The important message is that " every time read the labels on the food products specially for the carbohydrate, sugar and the fat contents and buy those with reasonable amount of carbohydrates and less than 4% sugar for a healthy diet.

For diabetic patients, rapidly digesting sugar specially sucrose and glucose can act as a sweet and slow poison. And hence it is a huge challenge for diabetic patients to refrain from taking these sugars in order to maintain the glucose level in their blood as normal as possible.

The irony of the fact is that it is very difficult to find such food, especially ready made sold in the market, which is totally free from sugar. So what diabetic patients can do? The suggestion is: do not buy such food which does not indicate what amount of sugar in term of % has been added. Food sold in the shop can carry from 60% down to 0.5% of sugar. Some sugary food may be highly tempting—use your will power and do not bring them home. For sweetening tea or coffee, a number of different sweeteners (saccharine, aspartame, sorbitol and few more) are available which are more suitable for diabetic patients. Normal soft drink ought to be replaced by diet drink.

As glucose is one of the most important components for our biological functions. Thus our life depends on the regular supply of glucose and more importantly to generate energy. It is the sustained high glucose level in the blood which leads to chronic complications. Therefore the question arises how to avoid this accumulation of large amount of free floating glucose molecules leading to diabetic complication, and also keep the continuous supply of glucose for biological activities and energy production.

Most glucose we obtain from external sources comes from food. Most food normally contains glucose in the form of carbohydrates. Once food is inside body the carbohydrates via various biochemical reactions are converted into glucose. Some carbohydrates are simple and broken down quickly while others such as starch (present on porridge, potato and rice) are broken down slowly.

Importance of Starchy Food

For diabetic patients, hence the advice is that they should consume the food which contains complex carbohydrates and in moderate quantity. There are a number of starchy foods available in which starch is present from moderate to high concentrations. These include porridge, rice, potatoes and pastas etc. They also contain most required nutrients including and the slow digesting starch into glucose which can be used by the body as fuel. Other good thing about the starchy food is that they naturally contain low fat and may contain high fiber which can additionally be helpful for keeping the bowel clear and prevent constipation.

It may be noted that although rice comes in the category of starchy food but to some people it can induce constipation. One reason for this is that, especially in Eastern countries, the rice is normally boiled and the excess water together with most starch is decanted out and thrown. For this reason the amount of such boiled rice is consumed in heavier quantity to satisfy the satiation. We recommend that the starch should be cooked with the amount of water that does not have to throw away any liquid and also eat less rice than normally needed.

Fat

Although fat does not play the same direct role in causing diabetic complications, it is equally important that patients should keep a control on fat intake. Fat is an important constituent of our food and for healthy life some right kind of fats are required to be consumed in right amount. Fat is a good source of energy and certain fat helps us in absorbing certain vitamins which our body cannot make. Despite these facts it is unhealthy to consume too much fat especially of wrong kind because it contributes to obesity, heart attack and stroke and even death—a complication more common among diabetics than non-diabetics.

Types of Fats

We consume different types of fat present in different food: **saturated** which is present in animal products such as milk, butter and meat and **trans-fat** which is present in processed food such as biscuits, cakes etc. They are produced from the heating of the various fat—and that is why it is recommended that left over oil from frying should not be used again and again (as it is done in some takeaways and restaurant) because the more it is heated the more trans-fat is generated and both types, saturated and trans-fat are unhealthy. Then there is unsaturated fat which is present in vegetables. Unsaturated fat are of two types: monounsaturated which are present in certain kind of oil such as obtained from mustard, canola and rape seed. Polyunsaturated oil are of two types Omega 3 which is present in fish oil such as salmon, mackerel, tuna and sardines and Omega 6 is present in corn, sunflower and ground nuts. All kinds of unsaturated fat are healthy and also involved in balancing cholesterol in body. For diabetic patients, especially if obese, it is important to stay away from the saturated fat and focus on consuming unsaturated types.

Alpha Lipoic Acid

Alpha lipoic acid (a fatty acid) is found naturally inside every cell in the body. Although can be synthesized within human bodies it can be obtained from the diet, such as red meat, spinach, broccoli, potatoes, yams, carrots, beet roots and yeast. When used as an oral dietary supplement, alpha lipoic acid has been shown to improve insulin sensitivity, glucose effectiveness and glucose disposal.

Proteins

It is well known that human body uses proteins to maintain normal biological functions, with leaving far less effects on the blood sugar levels . Most protein are available in products obtained from animals. Some plants like beans, , lentils and peas also contain protein. Plant products are very low in fat hence, when it comes to protein foods, one must always try to combine these protein containing plant foods with low fat in comparison to fat in animal products.

Recommended daily intake of protein is between 0.8 to 1 g/kg of the body weight. It is highly recommended to avoid most saturated fats and for this fishes are the best option to be eaten. Also non-fat milk and skimed milk on their own or products are useful. Eggs , skinless chicken, lean meats are also recommended

L-Arginine

L-arginine is an amino acid, one of the components of proteins, which is obtainable naturally from the diet. A number of diets such as fish, chicken, other meat, nuts and dairy products are rich in arginine (and other amino acids). L-arginine is sold in the market as food supplement as it is a good source of nitric oxide which triggers the relaxation of blood vessels to increase blood flow. Type 2 diabetic patients make reduced amount of nitric oxide and may suffer from erectile dysfunction.

Salt

Salt is an essential component of our food but too much salt has been shown to be health damaging particularly its role in increasing the blood pressure and subsequent stroke and heart disease. As diabetic patients are more prone to heart disease they should take special care to maintain good blood pressure and for that should take prescribed amount of salt per day.

Vitamin and Minerals

Vitamin D

Current studies show that vitamin D plays major roles in a number of human metabolic systems. Over 1,900 research papers have been published in this field and the work still progressing with a rapid speed. Vitamin D is present in liver, epidermis, thymus, small intestine and pancreas. Besides its classical roles in bone metabolism, vitamin D also plays roles in maintaining blood pressure, metabolic syndromes, kidney function, periodontal diseases, and inflammation and more importantly in cardiovascular risk. In diabetes vitamin D has roles in insulin resistance, hyperglycemia and glucose homeostasis.

Vitamin B1

Vitamin B1, also known as thiamine, is another vitamin claimed to have antidiabetic effects. It is a co-factor in carbohydrate metabolism and hence diabetic patients have its deficiency albeit with variable degrees. Thiamine deficiency is also linked with oxidative stress, inflammation and endothelial dysfunction and hence can lead to cardiovascular disease, atherosclerosis and the metabolic syndromes. Laboratory experiments have shown that this vitamin can prevent formation of harmful elements of glucose metabolism, but its use to prevent diabetic complications in man has yet to be proven.

Vitamin B$_{12}$

A link between long term metformin use and vitamin B$_{12}$ deficiency has been reported in certain cases of Type 2 diabetes and hence it may be wise to keep a check on certain symptoms of vitamin B$_{12}$ deficiency, if proven, can be treated easily by B$_{12}$ supplement.

17

Chromium

Chromium is a trivalent trace element that is nutritionally essential in humans, requiring 0.005–0.2 mg/day. This element has been found to play a role in maintaining insulin and carbohydrate metabolism.

Magnesium

Magnesium is a mineral involved in over 300 enzymatic reactions in the human body. There have been a number of observational studies undertaken on the role of magnesium in Type 2 diabetes. A meta-analysis was conducted to find a link between magnesium intake and risk of Type 2 diabetes. This element was found to be inversely associated with incidence of Type 2 diabetes. A daily oral dosage of this element in the treatment of diabetes is 2,500 mg per day.

Some Popular Food Supplements

Karela

It is a vegetable grown in tropical countries and contains bitter taste hence also known as bitter melon. The author does not understand why it is called "melon" as its appearance is very different from any "melon". The fruit and leaves of this cucumber family vegetable are used in different preparations and also as capsule of karela is suppose to have antidiabetic effect. However, sporadic evidence exists that this vegetable may have short time effect on lowering the blood glucose but more research is needed to confirm this claim. Even it may have some ability of reducing the blood sugar; the authors believe that you will require a huge amount of karela each time to benefit its effect.

Cinnamon

An ancient medicine used for various ailments contains an active ingredient, hydroxychalcone which is believed to enhance insulin action. Studies have been carried to find if it can reduce blood sugar alongside cholesterol and blood fat from Type 2 diabetic patients. Again conflicting results obtained and hence more research is needed to determine the valuable effect of cinnamon on diabetic patients.

Beta-Glucans

Beta-glucans are polysaccharides consisting of glucose monomers bonded by beta-linkages and is a source of soluble fiber commonly found in several plants, grains, algae and the cell walls of bacteria and fungi. In the treatment of diabetes, beta-glucans have been found to improve glucose, insulin and cholesterol levels in Type 2 diabetic patients.

Glucomannan

Glucomannan is a high-viscous water soluble polysaccharide fiber extracted from plants and widely used in Japan and Taiwan as a dietary staple. Consumption of 3.6 g of Konjac-glucomannan on daily basis can significantly lower total cholesterol, LDL-cholesterol, total/HDL cholesterol, ApoB and fasting glucose levels.

Psyllium

Psyllium from the plant Plantago ovatum (available at the counter) has been in use as food supplement, containing at least 67% soluble fiber. It has been mainly used as a bulk laxative, weight loss, and more recently for lowering blood sugar and cholesterol. Over the years, many clinical trials have found that ingestion of psyllium (averaging 10.2 g/day) significantly improved postprandial glucose and $HbA1_C$ levels in people with Type 2 diabetes.

17

Guar Gum

Guar gum, a galactomannan gum extracted from the endosperm of the guar bean, is mainly found in India and Pakistan and contains 75% soluble fiber. It has the ability to attenuate postprandial glucose levels as well as decrease cardiovascular risk.

Herbal Medications

Soluble Fiber

Dietary fibers have been categorized, based on its solubility in water, into viscous (soluble) and non-viscous (insoluble) forms. Several studies have discovered that soluble fiber has the ability to prevent and control diabetes and its complications by aiding in body weight reduction, improving insulin sensitivity, improving $HbA1_C$, and postprandial blood glucose levels, as well as serum lipid levels while research data surrounding the effect of insoluble fiber is still controversial.

Agaricus

Agaricus (*Agaricus blazei*) is a medicinal mushroom from the Agaricaceae family that has been shown to reduce insulin resistance. In an experimental trial for 12 weeks, participants taking agaricus showed significantly lower insulin resistance, as measured by homeostasis model assessment for insulin resistance (HOMA-IR) indices, versus the control group.

Ginseng

There are two types of ginseng that have been researched for their effects on diabetes: American Ginseng (Panax quinquefolius) and Asian/Korean ginseng (Panax ginseng). Although bearing the same name, these two ginsengs differ greatly when it comes to their overall effect on the body. American Ginseng has been found to possess more stress-relieving, digestion-improving and anti-aging qualities, while Asian/Korean ginseng improves blood flow, battles oxidative stress and improves energy.

Milk Thistle

Milk thistle (Silybum marianum) is a plant from the Asteraceae family that has been shown to lower fasting blood sugar as well as $HbA1_C$.

Gymnema

Gymnema (Gymnema sylvestre) is an herb native to the tropical forests of Southern and Central India that has traditionally been used to treat diabetes. An open label study of Type 2 diabetic patients demonstrated a significant reduction in fasting glucose, $HbA1_C$ and in glycosylated plasma proteins.

Salacia

Salacia (Salacia reticulata and oblonga) is a plant in the Celastraceae family reported to have antidiabetics properties in a limited numbers of clinical trials.

Prickly Pear Cactus

Prickly Pear Cactus, (Opuntia streptacantha) is a traditional plant grown in desert regions of Mexico as well as in certain regions of the Mediterranean countries and in the United States. Also known as noples, they are a dietary staple in Mexican culture and have been traditionally used to treat gastrointestinal problems such as ulcers, and it has recently gained interest with the medical community for its importance to treat Type 2 diabetes.

Pycnogenol

Pycnogenol is a water-soluble standardized extract from the plant French Maritime pine (Pinus maritima), shown to possess a number of benefits in those with Type 2 diabetes.

Maitake Mushroom

Maitake (Grifola frondosa) is a fungus mainly grown in Japan. The effect on diabetes has yet to be established in human trials, but a number of experiments on animal models have shown that maitake may be beneficial in the treatment and prevention of diabetes.

17

Special Guidance for Diabetic Patients

Now you have read the book (or about to go to read it), here are some special guidance and instructions for the patients suffering from diabetes.

Do You Have Diabetes or Anybody in the Family Suffering from It?

If you do not have diabetes, or any body suffering from it then you do not have to worry about it unless you are obese.

So What Do You Do if You Are Obese but Not Suffering from Diabetes?

As you may have learnt from this book, there is a close link between obesity and diabetes (specially of Type 2) and also you may have noted here that there is a gradual global increase in the incidence of diabetes, so much so it is taking the form of pandemic. Main reason of obesity is the sedentary life style and consumption of high caloric foods.

If You Are Obese and No History of Diabetes in the Family Still You Stand a Good Chance of Developing This Disease. So What Do You Do?

The best course of action you can take is to try to bring down your body mass index (BMI). BMI can be calculated by simply measuring your height and weight. Generally it is not easy to bring down the body fat and even those who try them hard (for example going to gym or cutting down the junk food intake) most cannot keep it going once the body weight goes down; they become relaxed and then the body weight starts going up again. So it is important to enter in those activities which can be adapted for long term, (perhaps for the entire life) and maintain the normal body weight. For the long term strategy to adapt, it is advisable to contact specialist in this field.

It is also important to note that, by getting rid of your obesity, you would be doing a great favor to your health for every obesity related complication such as heart attack, diabetes and diabetes related complication and many more disadvantages, including ugly look (possibly), run for your life, not

Diabetes: A Comprehensive Treatise for Patients and Care Givers,
by Shamim I. Ahmad and Khalid Imam. ©2014 Landes Bioscience.

able to participate in sports, feeling guilty to enjoy sugary and fatty meals, difficulty in finding comfortable seats in transportations (such as buses, trains and airplanes), and several more.

So:

If you are not obese and do not have family history of diabetes you do not have to worry, but it is wise to have your blood tested once in a year for hyperglycemia (high blood sugar) after the age of forty.

But What if You Have Diabetes in the Family?

In this case it is wise to have the blood glucose level checked on every six monthly basis starting from the age of around forty. It is very simple test and nothing to worry. Your doctor will take a pin-drop amount of blood to be tested using a glucometer and give you instant result. If the test is not confirmatory or you may have taken large amount of sugary food or drink just before test the result may appear misleading. In both the cases your doctor will advise you for further tests and in appropriate conditions.

You may also note that a large number of Type 2 diabetics have been found to have been diagnosed several years after its commencement and by then, in certain cases, it may be too late to reverse certain complications caused by diabetes.

So, If You Are Tested as Diabetes Positive?

You must enter in fighting back this disease and if you have read the book you must have noted:

1. Diabetes is a blessing in disguise;
2. Diabetes is a slow killer if not taken seriously;
3. Gaining knowledge about diabetes is itself a treatment.

Epilogue

To Fight Back Diabetes

Consider these three issues:

- Develop will power to fight back diabetes;
- Never feel ashamed of suffering from diabetes, specially when you understand that diabetes is one of the commonest disease and globally there are around 285 millions others who are suffering from it;
- Fight back any depressive feeling you may be encountering, thinking— why you? Consult your doctor if needed.

Gaining Knowledge about Diabetes

Like oxygen is vital for our survival, glucose is another very important constituent for life. Glucose performs many functions and its level in our body is intricately balanced (homeostasis) according to need. We produce glucose from our food (especially from carbohydrates) and utilize them as needed. Surplus glucose is stored in our body in various forms and is used up when needed. Brain is the most important organ to use glucose. The sugar levels in our blood (both diabetics and non-diabetics) fluctuate fast depending upon food intake, its carbohydrate type, your activity and the content and status of the underlying renal and liver function.

Our pancreas plays key role in producing insulin and this protein plays key roles in glucose homeostasis. Glucose floating in our blood must enter in cells to be utilized and insulin facilitates the entry of glucose in cells.

Different types of diabetes have now been identified but two most commons are—Type 1 and Type 2 diabetes. Your doctor will tell you which type of diabetes you are suffering from. If suffering from Type 1 (usually 5-7% of the total types) it is and more likely to be diagnosed when you are a teenager, this means your pancreas is no longer able to produce insulin and your doctor will immediately put you on insulin.

Type 2 diabetes (about 85% of the total types) is diagnosed usually after the age of 40 (but now is becoming more common in obese children). In this type of diabetes the pancreas may be producing subnormal amount of insulin. Additional or alternative reason for Type 2 diabetes is the cells developing resistance to allow insulin to transport glucose inside of the cell. This phenomenon is known as Insulin Resistance.

Managing Your Diabetes

Diabetes is an incurable disease and hence its control is vital to delay the onset of diabetes-associated complications as long as possible.

Type 1 diabetes: As said above the common treatment employed currently is the insulin injection and usually the patient is required to have it more than once daily. Also it is important to adjust the dose of insulin and keep a close watch (using a glucometer) on glucose level in the blood. Hypoglycemia (especially in very young children) must be carefully watched as well. Parents and school nurses must be trained to identify hypoglycemia in children and its aversion.

Type 2 diabetes: Although the molecular aspects of Type 2 diabetes is more complicated than Type 1 diabetes it is relatively easier to enter into its management. For management of this type of diabetes:

- Depending upon the level of hyperglycemia your doctor may advice you not to go for any antidiabetic drug yet. He almost certainly will ask you to cut down, as much as possible, on sugar and fat. Also go for healthier diet such as fresh vegetables and fruits, moderate amount of slow digesting carbohydrates, and fish and chicken as a source of protein.
- Exercise is the most important factor to maintain a good normoglycemic state. Which exercise you will enter into will depend upon a number of factors; cycling, swimming, brisk walking and even jogging is good exercise (see below).

- By developing will power, regular exercise, and healthy non-sugary and non-fatty food you can overpower you diabetes for considerably long time and avoid entering medication and into diabetic complications.

Ironically it is not possible to avoid antidiabetic treatment forever by the above practices only. Diabetic patients eventually will require antidiabetic medications.

Your doctor will keep a good record of your 3-6 monthly checks for your HbA1$_C$ and will advice when you have to start the medication.

Management of Drug Doses

It is crucial that right amount of anti-diabetic drugs (or insulin) are taken at right time on regular basis. Taking less than the amount required would be less effective to bring down hyperglycemia and taking higher dose of drug can lead to hypoglycemia. One major problem with higher dose of drug is that when a patient feels to have entered in hypoglycemia, to come out from it, if did not have proper guidance from the doctor, he may take larger amount of sugary material than needed. This in turn, although brings out the patient from hypoglycemic feelings (see above for symptoms of hypoglycemia) it can put him into hyperglycemic state for a long time. If this hypo/hyperglycemic conditions continue, when the next HbA1$_C$ is measured, the frequent hyperglycemic state may misguides the doctor that the antidiabetic drug is not working effectively. Hence the doctor may increase the dose further which is not right. Thus guidance from the doctor for managing hypoglycemia is very important.

As diabetes can lead to coronary heart disease, it is also valuable to add some kind of anticholesterol drug (such as a statin) in your medical regime. A doctor may also prescribe you low dose of aspirin and/or multivitamins as supplements.

THINGS MUST BE OVEREMPHASIZED

Stop smoking, receive care planning to meet your individual needs, attend diabetes educational courses, receive pediatric care for diabetic children and young adults, get special advice if planning to have a baby, see specialist diabetes healthcare professionals and get emotional and psychological support if required.

Diet for Diabetic Patients

Food selection is one of the most difficult tasks to combat diabetes. You have to select very carefully what food you can eat and what quantity so that only required amount of sugar enters in your biological system.

Another major problem that diabetic patients face is the inappropriate labeling of ready-made food. Most labels are fairly misleading, e.g., they may show carbohydrates, 20%, of which sugar is 10%. As we know sugar is a vague term that can include all different kinds of sugar such as glucose, lactose, fructose, mannitol, sorbitol, sucrose and many more. Not all sugars are harmful for diabetics, e.g., fructose and sorbitol are known to be safe but not glucose and lactose (present in milk). Taking yoghurt in place of milk is more appropriate as much of lactose is consumed by bacteria growing in milk and converting it into yoghurt. In one case the author noted a label saying "sugar free", but careful reading showed that it contained grape juice (high in glucose).

So what you do to organize your meal.

For breakfast: select from the following: egg, porridge, semi-skimmed milk or sugar-free yoghurt, whole meal bread, sugar-free peanut butter or diabetic jam, sugar free tea or coffee (can add artificial sugar recommended safer for diabetics) and low (1–3%) sugar added cereals.

For lunch: Fish or chicken, fresh vegetable and a small portion of fresh fruit, fresh salad as much as you feel.

If in habit of taking afternoon tea, you may take some salty snacks or low sugar biscuits with your tea.

For dinner: May take as shown for lunch but keep the amount less than your lunch and avoid fresh fruit. This is because when you are sleeping your glucose needed is much less than when you are active in the daytime.

Notes: Although this chapter suggests bringing about certain stringent controls in life style to delay the onset of diabetic complication you must also be sure that it should not occur at the cost of damaging the quality of life. For example it is a fact that if you consume sugar added food such as cake or sweets your sugar level will increase. To maintain life quality, in writer's opinion, occasionally if the sugar added food (e.g., a piece of birthday cake) in front of you become irresistible allow yourself to enjoy it—just make it sure it does not become a normal habit.

Exercise

Exercise is an important component of diabetic control. Its importance is reflected from the number of research and review papers published (11,683 papers until 2012; ref. PubMed) emphasizing the value of exercise in diabetes. Exercise not only leads to reduction in blood sugar levels but also many other valuable effects of exercise are noted for both non-diabetic persons and diabetic patients. Reduction in obesity is one important outcome of regular exercise and, as mentioned above, obesity and diabetes are closely linked. Other valuable effects of exercise are: improvement in cardiac function, psychological and behavioral changes and improvement in memory. Although much research has been carried out with patients suffering from Type 1 diabetes, exercise is equally important for those with Type 2 diabetes.

A good indication of fruitful exercise is a rise in heart rate. Brisk walking is better than slow walking and patients can increase the time they spend exercising gradually as their stamina improves. Current national recommendations for physical activity for adults are to do 150 minutes of moderate intensity activity (e.g., brisk walking), or 75 minutes of vigorous intensity activity per week and to perform strength training activities that involve all major muscle groups at least two days per week. However, some diabetic patients may be prohibited from exercise due to other health problems. Interestingly, for such cases a recent research has shown that 20 minutes of passive stretching (exercise) also can lower glucose levels in an at risk population.

Ranging from simple walk to strenuous exercise at gym, you can select one or more of these depending upon a number of factors including your age, blood pressure, heart condition and stamina. Here is author's suggestion:

Have the exercising facility such a tread mill machine or an exercise bicycle at home. This will give you the freedom to exercise any time you wish to and irrespective of weather condition prevailing. Also jogging or walking alone can be boring but at home during exercise you could develop other activities such as listening your favorite music, or reading a favorite novel etc.

Epilogue

Check-Ups for a Good Diabetes Care

On Daily Basis

Whether you are suffering from Type 1 or Type 2 diabetes, it is valuable to obtain a glucometer and check your blood sugar level fairly regularly, at least once a day. In the beginning of diabetes diagnosed you may have to check the blood sugar level for more than once a day, in the morning before breakfast, in the noon after meal and before and after exercise and once before going to bed. Within a week or two, after these data collection you will have a good idea about the daily fluctuation of your blood sugar levels and how the food and exercise, as well overnight sleep can influence the blood glucose level. Based on these data you may be able to manage your daily routine for better diabetes control.

At Every 3–6 Month Intervals

Check your $HbA1_C$. The result will suggest that during the last 3 or 6 month's period what has been your average glucose sugar level and this test is a must for each diabetic patient. A reading of 7.0% or less suggests a good diabetic control. Above 7.0% requires reconsideration of diabetic management.

Annual Check-Up

Annual check-ups should include, check-ups of blood pressure, weight, body mass index, peripheral pulses, eyes for retinopathy, feet for neuropathy and infections and kidney function tests by measuring serum creatinine and urine micro-albumin and finally fasting lipid profile.

Diabetes: A Comprehensive Treatise for Patients and Care Givers,
by Shamim I. Ahmad and Khalid Imam. ©2014 Landes Bioscience.

What Should Be Target Control

You should also know the desirable and optimum levels of good glucose, lipids and blood pressure control. By achieving these values you can prevent or delay the diabetes complications. These values are given below:

Glycemic Control
- Fasting glucose between 90–130 mg/dL (5.0–7.2 mmol/L)
- 2-hr post meal glucose less than 180 mg/dL (10 mmol/L)
- HbA1$_C$ ≤7%(53 mmol/mol)

However, normal glucose values for those who are not diabetic should be interpreted as follows:
- Fasting glucose less than 100 mg/dL (5.5 mmol/L)
- 2-hr post meal glucose less than 140 mg/dL (7.7 mmol/L)
- HbA1$_C$ ≤5.7% (38.8 mmol/mol)
- HbA1$_C$ between 5.7–6.4% (38.8–46.4 mmol/mol) consider prediabetics.

Cholesterol Control
- Total Cholesterol less than 200 md/dl (5.18 mmol/L)
- Triglycerides less than 150 mg/dL (3.88 mmol/L)
- LDL-Cholesterol less than 100 mg/dL (2.59 mmol/L)
- HDL-Cholesterol more than 40 in men (1.03 mmol/L), and more than 50 (1.29 mmol/L) in women

Blood Pressure Control
- Less than 130/80 mm Hg (strictly)

Kidney Function Test
- Serum creatinine around 1.0 mg/dL (91 micro mol/L) or less than 1.4 mg/dL (127 micrmol/L)
- Urine microalbumin less than 30 microgram/mg creatinine.

ALT (Liver Test)
- 30–40 iu

Diabetes: A Comprehensive Treatise for Patients and Care Givers,
by Shamim I. Ahmad and Khalid Imam. ©2014 Landes Bioscience.

Check List for Diabetes Caring

At Every Visit to Clinic, Check:
- Blood pressure
- Weight
- Body mass index
- Peripheral pulses

Have This Check-Up Done 3-6 Monthly:
- Glycosylated hemoglobin (HbA1$_C$)

Annual Check-Ups:
- Eyes check up for retinopathy
- Feet check up for neuropathy and infections
- Kidney functions check up: Serum creatinine and urine microalbumin
- Fasting lipid profile

Things Must To Do:
1. Stop smoking
2. Receive care planning to meet your individual needs
3. Attend diabetes educational courses
4. Receive pediatric care for diabetic children and adults
5. Get special advice if planning to have a baby
6. See specialist diabetes healthcare professionals
7. Get emotional and psychological support.
8. Keep HbA1$_C$ at desirable range
9. Keep fasting and random glucose with in optimum range
10. Get cholesterol levels with in target
11. Achieve desirable blood pressure goal

Diabetes: A Comprehensive Treatise for Patients and Care Givers,
by Shamim I. Ahmad and Khalid Imam. ©2014 Landes Bioscience.

Sources of Diabetes Education

There are many sources to get education material for diabetes awareness and self management, an integral part of diabetes management. These are enlisted below:

Internet

Internet is one of the best sources where a wealth of information can be found associated with all kinds of diabetes. In fact the amount of information both simple and complicated is so abundant that the reader may feel in a maze of information and do not know where to begin and where to end and their digestion becomes even more difficult. So it is advisable to collect and note those selected information easy to digest and follow.

Diabetes Societies

Diabetes Societies are the second best sources to get educated on diabetes. Some of these societies work at local levels, some at national levels and some at international levels.

Diabetes UK is one of the best and highly prestigious societies in Britain working in this field and doing a superb job. It was founded in 1934 by two famous personalities, RD Lawrence and HG Wells (both suffered from diabetes). There are two ways you can gather information about this society (a) by going through internet and (b) by joining the society as a member. The recommendation is that one should become a member. This will allow you to get a bimonthly magazine called BALANCE which can provide up-to-date information including latest research results and valuable guidance in the field.

The National Institute for Health and Clinical Excellence (NICE) is another organization geared to work for diabetes and it is useful to present below their views on delivering the diabetic programs.

- Improving knowledge, health beliefs, and lifestyle changes
- Improving patient outcomes, e.g. weight, hemoglobin $A1_C$, lipid levels, smoking, and psychosocial changes such as quality of life and levels of depression

- Improving levels of physical activity
- Reducing the need for, and potentially better targeting of, medication and other items such as blood testing strips
- All Primary Care Trusts (PCT) must commit to offering structured education programs to people with Type 2 diabetes from the point of diagnosis and as an ongoing part of their therapy in the long-term

Primary Care Trust

There are several different programs available across the United Kingdom, reflecting the different ways the PCTs have gone about establishing courses:

For Type 2 diabetics there are the Diabetes Education and Self-management for Ongoing and Newly Diagnosed (DESMOND) and X-PERT programs. For non-insulin users who have been recently diagnosed, the DESMOND course is highly recommended. For Type 1 diabetes the DAFNE course is ideal, whilst for those with long-standing Type 2 diabetes, the XPERT program is most appropriate.

For Type 1 diabetics there are the Dose Adjustment For Normal Eating (DAFNE), and Bournemouth Type 1 Intensive Education Program (BERTIE).

DiabetesSisters

Offers unique opportunities for women with diabetes and their loved ones to learn more about healthy living, gain the support needed to manage the emotional aspects of diabetes, and find ways to proactively advocate for women with diabetes. In fact, more than 123 million women around the world are living with diabetes; 11.1 million of those women are in the United States.

Below is presented the main areas covered in DAFNE course. Which diabetes course you enroll on depends on which type of diabetes you have, how recently you have been diagnosed, and your current management approach.

- Pathophysiology of diabetes
- Types of diabetes
- Metabolic control of diabetes and its monitoring
- The types, actions and duration of action of insulin preparations
- Nutritional food groups
- The concepts of carbohydrate portions and the glycemic index
- Adjusting short-acting insulin to the carbohydrate portions and glycemic index of a meal
- Avoiding weight gain
- Sweeteners and sugar substitutes

- Alcohol, insulin and diabetes
- Dose adjustment for snacks
- A step-wise approach to insulin dose adjustment
- How to deal with episodes of hyperglycemia
- Coping with intercurrent illness and adjusting the insulin dose when ill (using supplementary 10% and 20% of total daily insulin dose system with a ready reckoner for ease of use)
- The origin and symptoms of hypoglycemia
- Treating episodes of hypoglycemia
- Adjusting insulin dose following hypoglycemia
- Insulin adjustment for physical activity and exercise
- The purpose and content of the annual diabetic review
- Foot care
- Travelling with Type 1 diabetes
- Driving and Type 1 diabetes
- Pregnancy, contraception and Type 1 diabetes

How Does It Work?

- Those who attend the course are taught how to assess the carbohydrate portions (CPs) and glycemic index of the meals that they eat.
- A handy pocket book is provided to help with this, covering a wide range of commonly eaten foods, including trademarked brands.
- The patient's individual response to taking the recommended dose of insulin for the CPs eaten is assessed and the patient self-adjusts the amount of fast-acting insulin they take for a given quantity of CPs, and according to their preprandial capillary blood glucose. This helps to improve glycemic control, and encourages the patient to analyze, rather than just record, their capillary blood glucose measurements.

The patient is given a step-wise approach on how to adjust both fast-acting and long-acting insulin where glycemic control can be improved.

App IV

A very large number of books on various aspects of diabetes have been published, by various authors and various publications. Below are some suggested books and most of them may be available on mail order or through your bookshops.

Diabetes UK, one of the oldest and largest and organization based in United Kingdom have published a large number of leaflet, magazines and guide books on this subject. Their address is: Macleod House, 10 Parkway, London NW1 7AA. Telephone: 020 7424 1000 Email: info@diabetes.org.uk Website: www.diabetes.org.uk

Below we produce a list which can be requested from the above organization.

    a. Type 2 diabetes: What you need to know
    b. Diabetes care and you
    c. Food and diabetes: How to get it right
    d. Healthy eating for South Asian community – in Hindi and English combined
    e. Catalogue: *Information for people with diabetes and healthcare professionals*
    f. Diabetes UK (catalogue) Spring Summer 2011.

Bibliography

1. http://www.ncbi.nlm.nih.gov/pubmed/11953758
2. http://www.japi.org/special_issue_april_2011/01_Diabetic_History.pdf
3. Dobson M. "Nature of the urine in diabetes". Medical Observations and Inquiries 5. 1776:298-310.
4. Medvei, Victor Cornelius. The History of Clinical Endocrinology. Carnforth, Lancs., UK: Parthenon Pub. Group, 1993:23-34. ISBN 1-85070-427-9.
5. King H et al. Global burden of diabetes, 1995-2025: prevalence, numerical estimates, and projections. Diabetes care 1998; 21(9):1414-1431.
6. IDF Diabetes Atlas, 4th edition; 2009.
7. Yang W, Lu J, Weng J, et al. Prevalence of diabetes among men and women in China. N Engl J Med 2010; 362(12):1090-101.
8. Gale EA. The rise of childhood type 1 diabetes in the 20th century. Diabetes 2002; 51(12):3353-61.
9. Fox CS, Pencina MJ, Meigs JB, et al. Trends in the incidence of type 2 diabetes mellitus from the 1970s to the 1990s: the Framingham Heart Study. Circulation. 2006; 113(25):2914-8.
10. Wild S et al. Diabetes Care. 2004; 27:1047-1053.

Diabetes: A Comprehensive Treatise for Patients and Care Givers,
by Shamim I. Ahmad and Khalid Imam. ©2013 Landes Bioscience.

11. WHO. 2011; Available from: www.who.int/diabetes/en/.
12. Engelgau MM, Geiss LS, Saaddine JB, et al. The evolving diabetes burden in the United States. Ann Intern Med 2004; 140:945.
13. Steven RG. Emerging Epidemic: Diabetes in Older Adults: Demography, Economic Impact, and Pathophysiology. Diabetes Spectrum 2006; 19:221-228.
14. Frequency of metabolic syndrome in type 2 diabetics. Imam K, Shahid K, Hassan A, Alavi Z. JPMA 2007; 55:239-242.
15. DeFronzo RA, Ferrannini E. Insulin resistance. A multifaceted syndrome responsible for NIDDM, obesity, hypertension, dyslipidemia, and athero-sclerotic cardiovascular disease. Diabetes Care 1991; 14:173.
16. Grundy SM. Metabolic syndrome: a multiplex cardiovascular risk factor. J Clin Endocrinol Metab 2007; 92(2):399-404.
17. Mokdad AH, Ford ES, Bowman BA, et al. Prevalence of obesity, diabetes, and obesity-related health risk factors. JAMA 2003; 289(1):76-9.
18. Klein S, Burke LE, Bray GA, et al. Clinical implications of obesity with specific focus on cardiovascular disease: a statement for professionals from the American Heart Association Council on Nutrition, Physical Activity, and Metabolism. Circulation 2004; 110(18):2952-67.
19. Ritenbaugh C, Goodby CS. Beyond the thrifty gene: metabolic implications of prehistoric migration into the New World. Med Anthropol 1989; 11:227-223.
20. Classification and Diagnosis of Diabetes. Clinical practice recommendation. Diabetes Care 2011; 26(Suppl 1):S11-13.
21. Eisenbarth GS. Classification, diagnostic tests and pathogenesis. In: Becker KL, ed. Principles and Practice of Endocrinology and Metabolism. Philadelphia: Lippincott Williams and Wilkins, 2001:1315-1319.
22. Gat-Yablonski G, Shalitin S, Phillip M. Maturity onset diabetes of the young--review. Pediatr Endocrinol Rev 2006; 3(Suppl 3):514-20.
23. Holmkvist J, Almgren P, Lyssenko V, et al. Common variants in maturity-onset diabetes of the young genes and future risk of type 2 diabetes. Diabetes 2008; 57(6):1738-44.
24. Vehik K, Hamman RF, Lezotte D, et al. Increasing incidence of type 1 diabetes in 0- to 17-year-old Colorado youth. Diabetes Care 2007; 30(3):503-9.
25. Gale EA. The rise of childhood type 1 diabetes in the 20th century. Diabetes 2002; 51(12):3353-61.
26. Todd JA, Walker NM, Cooper JD, et al. Robust associations of four new chromosome regions from genome-wide analyses of type 1 diabetes. Nat Genet 2007; 39(7):857-64.
27. Neel JV. "Diabetes mellitus: a "thrifty" genotype rendered detrimental by "progress"?". Am J Hum Genet 1962; 14: 353-62.
28. DeFronzo RA: Pathogenesis of type 2 diabetes mellitus. Med Clin North Am 2004; 88:787-835.
29. Atkinson MA, Maclaren NK. The pathogenesis of insulin-dependent diabetes mellitus. N Engl J Med 1994; 331:1428.
30. Niskanen LK, Tuomi T, Karjalainen J, et al. GAD antibodies in NIDDM. Ten-year follow-up from the diagnosis. Diabetes Care 1995 Dec; 18(12):1557-65.

31. Purnell JQ, Dev RK, Steffes MW, et al. Relationship of family history of type
     2 diabetes, hypoglycemia, and autoantibodies to weight gain and lipids with
     intensive and conventional therapy in the Diabetes Control and Complications
     Trial. Diabetes 2003; 52(10):2623-9.

32. Redondo MJ, Rewers M, Yu L, et al. Genetic determination of islet cell autoim-
     munity in monozygotic twin, dizygotic twin, and non-twin siblings of patients
     with type 1 diabetes: prospective twin study. BMJ 1999; 318(7185):698-702.

33. Beck-Nielsen H, Groop LC. Metabolic and genetic characterization of pre-
     diabetic states. Sequence of events leading to non-insulin-dependent diabetes
     mellitus. J Clin Invest 1994; 94:1714.

34. Chen M, Bergman RN, Pacini G, Porte D. Pathogenesis of age-related glucose
     intolerance in man: insulin resistance and decreased beta-cell function. J Clin
     Endocrinol Metab 1985; 60:13-20.

35. Jack LL, Irl BH, Kevin AP, et al. Targeting β-cell function early in the course
     of therapy for type 2 diabetes mellitus. JCEM 2010; 95:4206-4216.

36. Yki-Ja"rvinen H. Glucose toxicity. Endocr Rev 1992; 13:415-431.

37. Robertson RP, Olson LK, Zhang HJ. Differentiating glucose toxicity from
     glucose desensitization: a new message from the insulin gene. Diabetes 1994;
     43:1085-1089.

38. Olefsky J, Farquhar JW, Reaven G. Relationship between fasting plasma insulin
     level and resistance to insulin-mediated glucose uptake in normal and diabetic
     subjects. Diabetes 1973; 22:507-513.

39. Chia CW, Egan JM. Incretin-based therapies in type 2 diabetes mellitus. J
     Clin Endocrinol Metab 2008; 93(10):3703-3716.

40. Toft-Nielsen MB, Damholt MB, Madsbad S, et al. Determinants of the im-
     paired secretion of glucagon-like peptide-1 in type 2 diabetic patients. J Clin
     Endocrinol Metab 2001; 86:3717-3723.

41. Farilla L, Bulotta A, Hirshberg B, et al. Glucagon-like peptide 1 inhibits cell
     apoptosis and improves glucose responsiveness of freshly isolated human islets.
     Endocrinology 2003; 144:5149-5158.

42. Brubaker PL, Drucker DJ. Minireview: glucagon-like peptides regulate cell
     proliferation and apoptosis in the pancreas, gut, and central nervous system.
     Endocrinology 2004; 145:2653-2659.

43. Zhou YP, Grill V. Long term exposure to fatty acids and ketones inhibits beta
     cell functions in human pancreatic islets of Langerhans. J Clin Endocrinol
     Metab 1995; 80:1584-1590.

44. Rothman DL, Magnusson I, Cline G, et al. Decreased muscle glucose trans-
     port/phosphorylation is an early defect in the pathogenesis of non-insulin-de-
     pendent diabetes mellitus. Proc Natl Acad Sci USA 1995; 92:983.

45. Boden G, Chen X. Effects of fat on glucose uptake and utilization in patients
     with non-insulin-dependent diabetes. J Clin Invest 1995; 96:1261.

46. Rasouli N, Kern PA, Adipocytokines and the metabolic complications of
     obesity. J Clin Endocrinol Metab 2008; 93(11 Suppl 1):S64-S73.

47. Dardeno TA, Chou SH, Moon HS et al. Leptin in human physiolody and
     therapeutics. Front Neuroendocrine 2010; 31(3):377-393.

48. Christian M, Parker MG. The engineering of brown fat. J Molec Cell Biol
     2010; 2(1):23-25.

49. Lee JM, Kim SR, Yoo SJ, et al. The relationship between adipokines, metabolic parameters and insulin resistance in patients with metabolic syndrome and type 2 diabetes. J Int Med Res 2009; 37:1803-1812.

50. Knip M, Veijola R, Virtanen SM et all. Enviromental triggers and determinants of type 1 diabetes. Diabetes 2005; 54(Suppl 2):S125-S136.

51. Pociot F, Akolkar B, Concannon P et al. Genetics of type 1 diabetes: What is next? Diabetes 2010; 59(7):1561-1571.

52. Cnop M, Welsh N, Jonas JC et al. Mechanism of pancreatic beta-cell death in type 1 and type 2 diabetes: many differences, few similarities. Diabetes 2005; 54(Suppl 2):S97-S107.

53. Mathieu C, Gysemans C, Giulietti A, Bouillon R. Vitamin D and diabetes. Diabetologia 2005; 48(7)1247-57.

54. Bin-Abbas BS, Jabari MA, Issa SD, et al. Vitamin D levels in Saudi Children with type 1 diabetes.Saudi Med J. 2011;32(6):589-92.

55. Flodstorm M, Maday A, Balakrishna D, et al. Target cell defense prevents the development of diabetes after viral infection. Nat Immunol 2002; 3(4):373-382.

56. Bach JF. The effect of infections on susceptibility to autoimmune and allergic diseases. N Engl J Med 2002; 347(12):911-920.

57. Greenspan FS, Gardner DG. Clinical Features of Diabetes mellitus; 641-43. Basic and Clinical Endocrinology (6th ed.) McGraw Hill, 2001.

58. Cryer PE et al. Evaluation and management of adult hypoglycemic disorders: an Endocrine Society Clinical Practice Guideline. J Clin Endocrinol Metab. 2009; 94(3):709-28.

59. Stephen JM, Maxine AP. Diabetes mellitus and hypoglycemia. Current Medical Diagnosis and Treatment 2010. McGraw Hill, 2010:1079-1122.

60. Classification and Diagnosis of Diabetes: Clinical Practice Recommendation. Diabetes Care 2011; 26(Suppl 1):S11-S13.

61. Nathan DM, Singer DE, Hurxthal K. The clinical information value of the glycosylated hemoglobin assay. N Engl J Med 1984; 310(6):341-6.

62. Nathan DM, Kuenen J, Borg R, et al.Translating the A1C Assay into Estimated Average Glucose Values. Diabetes Care 2008.

63. Panzer S; Kronik G; Lechner K, et al. Glycosylated hemoglobins (GHb): an index of red cell survival. Blood 1982; 59(6):1348-50.

64. Fong DS, Aiello LP, Ferris FL 3rd, Klein R. Diabetic retinopathy. Diabetes Care 2004; 27:2540-2553.

65. Gross JL, de Azevedo MJ, Silveiro SP, et al. Diabetic nephropathy: diagnosis, prevention, and treatment. Diabetes Care 2005; 28:164-176.

66. Boulton AJ, Vinik AI, Arezzo JC, et al. Diabetic neuropathies: a statement by the American Diabetes Association. Diabetes Care 2005; 28:956-962.

67. Boyle PJ. Diabetes mellitus and macrovascular disease: mechanisms and mediators. Am J Med 2007; 120:S12-S17.

68. Almdal T, Scharling H, Jensen JS, Vestergaard H. The independent effect of type 2 diabetes mellitus on ischemic heart disease, stroke, and death: a population-based study of 13,000 men and women with 20 years of follow-up. Arch Intern Med 2004; 164:1422-1426.

69. Colhoun HM, Betteridge DJ, Durrington PN, et al. Primary prevention of cardiovascular disease with atorvastatin in type 2 diabetes in the Collaborative Atorvastatin Diabetes Study (CARDS): multicentre randomised placebo-controlled trial. Lancet 2004; 364:685-696.

70. Kitabchi AE, Umpierrez GE, Murphy MB, et al. Management of hyperglycemic crises in patients with diabetes (Technical Review). Diabetes Care 2001; 24:131-153. Screening for type 2 diabetes. Diabetes Care 2011; 34(Suppl 1):S4.

71. Norris SL, Zhang X, Avenell A, et al. Long-term effectiveness of lifestyle and behavioral weight loss interventions in adults with type 2 diabetes: a meta-analysis. Am J Med 2004; 117:762.

72. Uusitupa M, Laitinen J, Siitonen O, et al. The maintenance of improved metabolic control after intensified diet therapy in recent type 2 diabetes. Diabetes Res Clin Pract 1993; 19:227.

73. Karra E et al. Mechanisms facilitating weight loss and resolution of type 2 diabetes following bariatric surgery. Trends Endocrinol Metab 2010. [Epub ahead of print] [PMID: 20133150].

74. Ismail K, Winkley K, Rabe-Hesketh S. Systematic review and meta-analysis of randomised controlled trials of psychological interventions to improve glycaemic control in patients with type 2 diabetes. Lancet 2004; 363:1589.

75. Nathan DM, Buse JB, Davidson MB, et al. Medical Management of Hyperglycemia in Type 2 Diabetes: A Consensus Algorithm for the Initiation and Adjustment of Therapy: A consensus statement of the American Diabetes Association and the European Association for the Study of Diabetes. Diabetes Care 2009; 32:193.

76. Monami M, Lamanna C, Marchionni N, Mannucci E. Comparison of different drugs as add-on treatments to metformin in type 2 diabetes: A meta-analysis. Diabetes Res Clin Pract 2008; 79:196.

77. Effect of intensive blood-glucose control with metformin on complications in overweight patients with type 2 diabetes (UKPDS 34). UK Prospective Diabetes Study (UKPDS) Group. Lancet 1998; 352(9131):854-65.

78. Bolen S, Feldman L, Vassy J, et al. Systematic review: comparative effectiveness and safety of oral medications for type 2 diabetes mellitus. Ann Intern Med 2007; 147:386.

79. Phung OJ, Scholle JM, Talwar M, Coleman CI. Effect of noninsulin antidiabetic drugs added to metformin therapy on glycemic control, weight gain, and hypoglycemia in type 2 diabetes. JAMA 2010; 303:1410.

80. Goldstein BJ. Clinical Translation of "A Diabetes Outcome Progression Trial": ADOPT Appropriate Combination Oral Therapies in Type 2 Diabetes. J Clin Endocrinol Metab 2007; 92:1226.

81. Kipnes MS, Krosnick A, Rendell MS, et al. Pioglitazone hydrochloride in combination with sulfonylurea therapy improves glycemic control in patients with type 2 diabetes mellitus: a randomized, placebo-controlled study. Am J Med 2001; 111:10.

82. Chiasson JL, Josse RG, Hunt JA, et al. The efficacy of acarbose in the treatment of patients with non-insulin-dependent diabetes mellitus. A multicenter controlled clinical trial. Ann Intern Med 1994; 121:928.

83. Schernthaner G, Matthews DR, Charbonnel B, et al. Efficacy and safety of pioglitazone versus metformin in patients with type 2 diabetes mellitus: a double-blind, randomized trial. J Clin Endocrinol Metab 2004; 89:6068.

84. Kahn SE, Haffner SM, Heise MA, et al. Glycemic durability of rosiglitazone, metformin, or glyburide monotherapy. N Engl J Med 2006; 355:2427.

85. Glucose transporters in human renal proximal tubular cells isolated from the urine of patients with non-insulin dependent diabetes. Diabetes 2005; 54:3427-3434.

86. Effect of intensive insulin therapy on residual beta-cell function in patients with type 1 diabetes in the diabetes control and complications trial. A randomized, controlled trial. The Diabetes Control and Complications Trial Research Group. Ann Intern Med 1998; 128:517.

87. Bolli GB, Kerr D, Thomas R, et al. Comparison of a multiple daily insulin injection regimen (basal once-daily glargine plus mealtime lispro) and continuous subcutaneous insulin infusion (lispro) in type 1 diabetes: a randomized open parallel multicenter study. Diabetes Care 2009; 32:1170.

88. Singh SR, Ahmad F, Lal A, et al. Efficacy and safety of insulin analogues for the management of diabetes mellitus: a meta-analysis. CMAJ 2009; 180:385.

89. Havelund S, Plum A, Ribel U, et al. The mechanism of protraction of insulin detemir, a long-acting, acylated analog of human insulin. Pharm Res 2004; 21:1498.

90. ADVANCE Collaborative Group; Patel A et al. Intensive blood glucose control and vascular outcomes in patients with type 2 diabetes. N Engl J Med. 2008; 358(24):2560-72.

91. Westermark P, Johnson KH, O'Brien TD. Islet amyloid polypeptide — a novel controversy in diabetes research. Diabetologia 1992; 35:297.

92. Makimattila S, Fineman MS, Yki-Jarvinen H. Deficiency of total and non-glycosylated amylin in plasma characterizes subjects with impaired glucose tolerance and type 2 diabetes. J Clin Endocrinol Metab 2000; 85:2822.

93. Nauck MA, Homberger E, Siegel EG, et al. Incretin effects of increasing glucose loads in man calculated from venous insulin and C-peptide responses. J Clin Endocrinol Metab 1986; 63:492-498.

94. Egan JM, Bulotta A, Hui H, Perfetti R. GLP-1 receptor agonists are growth and differentiation factors for pancreatic islet cells. Diabetes Metab Res Rev 2003; 19:115-123.

95. Verdich C, Flint A, Gutzwiller JP, et al. A meta-analysis of the effect of glucagon-like peptide-1 (7–36) amide on ad libitum energy intake in humans. J Clin Endocrinol Metab 2001; 86:4382-4389.

96. van Bon AC, Brouwer TB, von Basum G, et al. Future acceptance of an artificial pancreas in adults with type 1 diabetes. Diabetes Technol Ther 2011; Apr 19.

97. Ojo AO, Meier-Kriesche HU, Hanson JA, et al. The impact of simultaneous pancreas-kidney transplantation on long-term patient survival. Transplantation 2001; 71:82-90.

98. Larsen. Pancreas transplantation: indications and consequences. Endocr Rev JL 2004; 25:919-946.

99. Alejandro R, Barton FB, Hering BJ, Wease S. Update from the Collaborative Islet Transplant Registry. Transplantation 2008; 86:1783-1788.

100. Nagaya M, Katsuta H, Kaneto H, et al. Adult mouse intrahepatic biliary epithelial cells induced in vitro to become insulin-producing cells. J Endocrinol 2009; 201:37-47.

101. Expert Committee on the Diagnosis and Classification of Diabetes Mellitus: Report of the expert committee on the diagnosis and classification of diabetes mellitus. Diabetes Care 26 2003; (Suppl 1):S5-S20.

102. American Diabetes Association: Gestational diabetes mellitus (Position Statement). Diabetes Care 27 2004; (Suppl 1):S88-S90.

103. Metzger BE, Lowe LP, Dyer AR, et al. Hyperglycemia and adverse pregnancy outcomes. N Engl J Med 2008; 358:1991.

104. Carpenter MW, Coustan D. Criteria for screening tests for gestational diabetes. Am J Obstet Gynecol 1982; 144:768-773.

105. Hillier TA, Pedula KL, Vesco KK, et al. Excess gestational weight gain: modifying fetal macrosomia risk associated with maternal glucose. Obstet Gynecol 2008; 112:1007.

106. Sheffield JS, Butler-Koster EL, Casey BM, et al. Maternal diabetes mellitus and infant malformations. Obstet Gynecol 2002; 100:925-930.

107. American College of Obstetrics and Gynecologists Committee on Practice Bulletins—Obstetrics: Gestational diabetes. Obstet Gynecol 2001; 98:525-538.

108. Dudley, DJ. Diabetic-associated stillbirth: incidence, pathophysiology, and prevention. Obstet Gynecol Clin North Am 2007; 34:293.

109. Landon MB, Hauth JC, Leveno KJ, et al. Maternal and perinatal outcomes associated with a trial of labor after prior cesarean delivery. N Engl J Med 2004; 351:2581.

110. Landon MB, Spong CY, Thom E, et al. A multicenter, randomized trial of treatment for mild gestational diabetes. N Engl J Med 2009; 361:1339.

111. American College of Obstetricians and Gynecologists. Gestational Diabetes. ACOG practice bulletin #30, American College of Obstetricians and Gynecologists. Washington, DC, 2001.

112. Oken E, Ning Y, Rifas-Shiman SL, et al. Associations of physical activity and inactivity before and during pregnancy with glucose tolerance. Obstet Gynecol 2006; 108:1200.

113. Kremer CJ, Duff P. Glyburide for the treatment of gestational diabetes. Am J Obstet Gynecol 2004; 190:1438.

114. Conway DL, Gonzales O, Skiver D. Use of glyburide for the treatment of gestational diabetes: the San Antonio experience. J Matern Fetal Neonatal Med 2004; 15:51.

115. Rochon M, Rand L, Roth L, Gaddipati S. Glyburide for the management of gestational diabetes: risk factors predictive of failure and associated pregnancy outcomes. Am J Obstet Gynecol 2006; 195:1090.

116. Moore LE, Clokey D, Rappaport VJ, Curet LB. Metformin compared with glyburide in gestational diabetes: a randomized controlled trial. Obstet Gynecol 2010; 115:55.

117. Schmidt MI, et al. Gestational diabetes mellitus diagnosed with a 2-h 75-g oral glucose tolerance test and adverse pregnancy outcomes. Diabetes Care 2001; 24:1151-1155.

118. Dodd JM, Crowther CA, Antoniou G, et al. Screening for gestational diabetes: The effect of varying blood glucose definitions in the prediction of adverse maternal and infant health outcomes. Aust N Z J Obstet Gynaecol 2007; 47:307.

119. Metzger BE, Lowe LP, Dyer AR, et al. Hyperglycemia and adverse pregnancy outcomes. N Engl J Med 2008; 358:1991.

120. Pettitt DJ, Knowler WC, Baird HR, Bennett PH. Gestational diabetes: infant and maternal complications of pregnancy in relation to third-trimester glucose tolerance in the Pima Indians. Diabetes Care 1980; 3:458.

121. Gabbe SG, Graves C. Management of diabetes mellitus complicating pregnancy. Obstet Gynecol 2003; 102:857-868.

122. Jensen DM, Korsholm L, Ovesen P, et al. Adverse pregnancy outcome in women with mild glucose intolerance: is there a clinically meaningful threshold value for glucose? Acta Obstet Gynecol Scand 2008; 87:59.

123. Sheffield JS, Butler-Koster EL, Casey BM, et al. Maternal diabetes mellitus and infant malformations. Obstet Gynecol 2002; 100:925.

124. Bellamy L, Casas JP, Hingorani AD, Williams D. Type 2 diabetes mellitus after gestational diabetes: a systematic review and meta-analysis. Lancet 2009; 373:1773.

125. Kim C, Newton KM, Knopp RH. Gestational diabetes and the incidence of type 2 diabetes. Diabetes Care 2002; 25:1862-1868.

126. Bilous R. Diabetes (Understanding). Family Doctor Books 2009.

127. Cheyette C. Balolia Y. Carbs & Cals: Count Your Carbs & Calories with Over 1,700 Food & Drink Photos! 2013.

128. Becker G. Type 2 Diabetes: the First Year - An Essential Guide for the Newly Diagnosed. Patient-expert Guides 2004.

129. Holford P. Say No to Diabetes: 10 Secrets to Preventing and Reversing Diabetes. 2011.

130. Govindji A, Suthering J. The Diabetes Weight Loss Diet. 2008.

131. Ruhl J. Blood Sugar 101: What They Don't Tell You About Diabetes. 2008.

132. Hanas R. Type 1 Diabetes in Children Adolescents. 2012.

133. Besser R. Diabetes Through the Looking Glass: Seeing Diabetes from Your Child's Perspective: A Book for Parents of Children. 2009.

134. Ezrin C, Kowalski RE. The Type 2 Diabetes Diet Book, Fourth Edition. 2011.

135. Thompson AW, Govindji A. Healthy Eating for Diabetes: In Association with Diabetes UK (Healthy Eating Series). 2009.

136. Bernstein RK. Dr Bernstein's Diabetes Solution: A Complete Guide To Achieving Normal Blood Sugars, 4th Edition. 2011.

137. Whettem E. Diabetes. Nursing and Health Survival Guides. 2012.

138. Nash J. Diabetes and Wellbeing: Managing the Psychological and Emotional Challenges of Diabetes Types 1 and 2. 2013.

139. Hyman M. The Blood Sugar Solution: The Bestselling Programme for Preventing Diabetes, Losing Weight and Feeling Great. 2012.

140. Miller S. Five Simple Steps To Cut Down On Sugar (Healthy & Tasty Series) by Fitness Books - Fast Fitness by SMGC Publishing (2012), 2013.

141. Ahmad SI, ed. Diabetes: An Old Disease, a New Insight. Austin/New York: Landes Bioscience/Springer Science+Business Media, 2012.

9 781570 597756